IMPOSSIBLE THINGS

21
IMPOSSIBLE THINGS

Nury Vittachi

Illustrated by Step Cheung

WS Education

Contents

Introduction
21 Impossible Things
p. vii

1 Reality is Not What it Seems to Be
p. 1

2 Seeing What Isn't There
p. 9

3 A World of Dust
p. 17

4 Reality is a Complicated Place
p. 25

5 Stepping Beyond Reality
p. 33

6 An Invisible Dimension Exists
p. 41

7 A Football in Place
p. 49

8 The Mysteriously Heavy Feather
p. 57

9 Keeping Body and Soul Together
p. 65

10 You Can Move an Elephant
p. 73

11 Invisible Hands, Everywhere
p. 81

12 Is This the Real Life?
p. 89

13
The Time Traveller

p. 97

14
Time for a Change

p. 105

15
A Ripple in Reality

p. 113

16
A World Made of Nothing

p. 121

17
Reality Disappears

p. 129

18
The Mystery of Free Will

p. 137

19
The Most Famous Cat in Science

p. 145

20
The Cursed Scientist

p. 153

21
Twins Who are a Universe Apart

p. 161

Epilogue
The Journey is Just Beginning

p. 168

22
Bonus tale: The Do-It-Yourself Universe

p. 179

Introduction:

21 Impossible Things

Let's start with a quiz!

But don't worry, there are only two questions, and they are pretty straightforward:

1) Is the world real?

2) How long is one second — is it one second long?

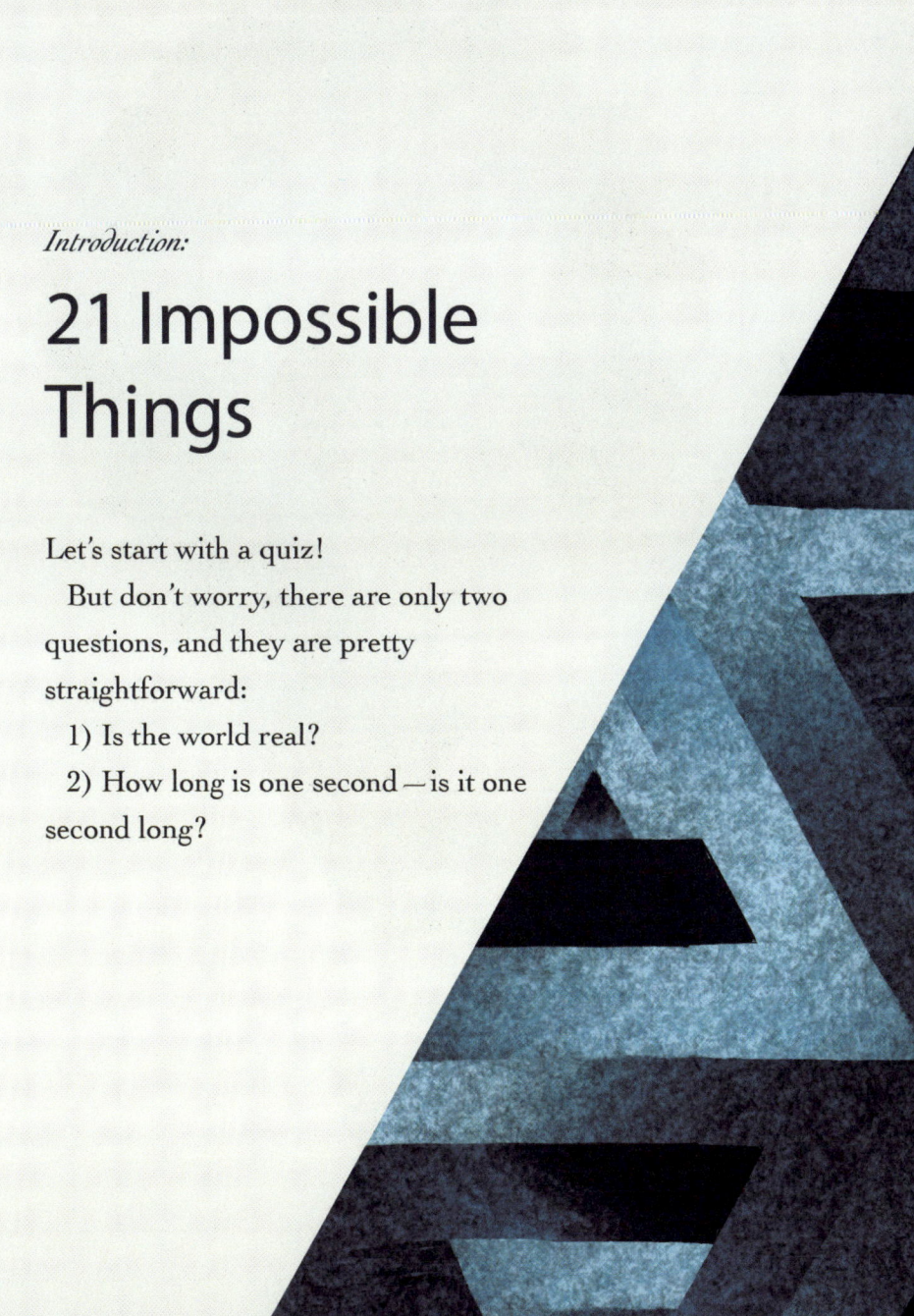

To most people, the answers are "Yes" and "Yes", and they would find it impossible to accept that they could be anything else.

And *that's* where it gets interesting.

Those are the logical answers. They are the rational answers — in fact, they are the only answers that make sense in our modern, scientific way of thinking.

And yet, after the development of two areas of science, we know that both answers are wrong.

Now I almost wrote "two new areas of science". But we too often forget that the underlying ideas are among the oldest surviving notions of history. We find them in plenty of ancient texts; in books more related to philosophy and religion.

Humanity has been on a long and astonishing journey to discover the true nature of reality, and it's one we can embark on in this book.

The Quantum Physics Puzzle

We start by dipping our toes gently into science.

People associate the major branches of science — biology, chemistry, physics, and so on — with data and logic. We watch processes happen, we measure things, and we make sure we understand what's going on. The focus is on a rational look at physical processes.

Quantum physics is exactly the same as other sciences in most ways, yet quite different in other ways. More than 100 years after it was developed, key aspects of it continue to defy "rational" explanation — even after endless experiments and vast amounts of data. One of its founders spotted this right at the beginning and said it seemed more like mysticism!

Despite this, it works in the physical world, and we use its principles every day. Indeed, quantum physics has become a science that underlies all other physical sciences.

A Second Reality

Some of the conclusions to which it leads us are extraordinary.

For one, they indicate that our current reality is supported by another, quite different reality: one which is invisible, yet more basic and more fundamental than ours. Quantum scientists say that that mysterious other layer is "real" reality. And that means that this reality, the one in which we live in, appears to be some sort of projection.

Strange, right?

Our Presence Changes Things

That's just the beginning of the oddness. Quantum physics also moves us away from what is called an "objective" view of reality to its opposite. Objective means that things are the same whether we are looking at them or not. Instead, quantum physics leans towards a "subjective" view of reality. That means that the presence of the observer has a direct effect on the thing being observed, at every level.

This is an intriguing point because in some ways it feels quite wrong, yet it feels naturally right, too. It moves us towards a very different understanding of the relationship between thinking, conscious human beings, and the physical world in which we live in.

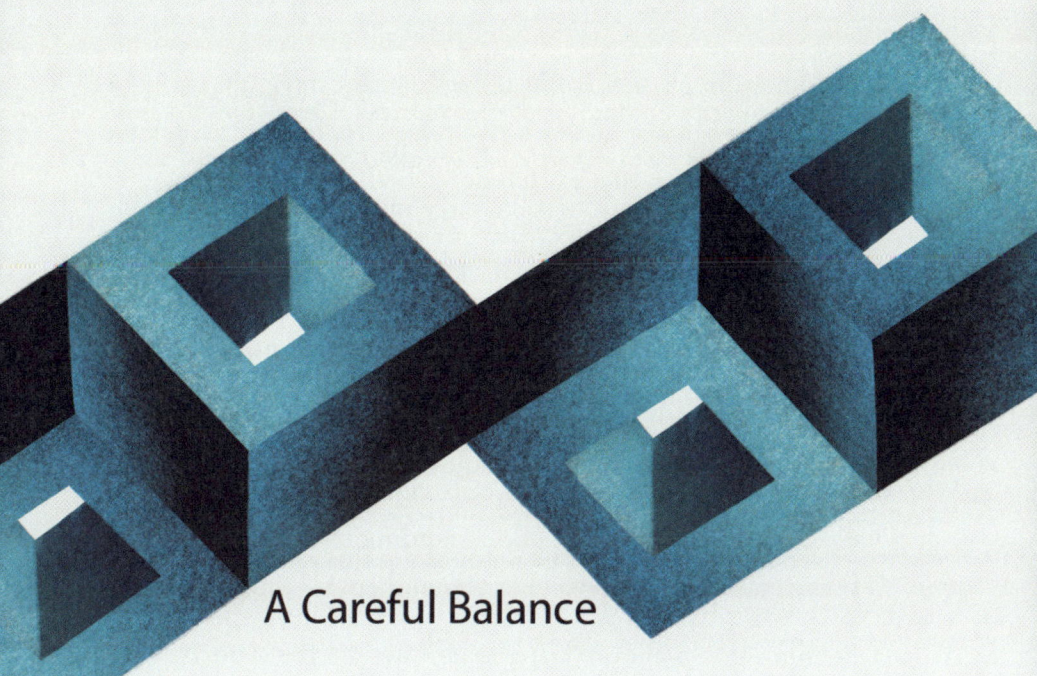

A Careful Balance

Some people like to take this notion to extremes, and have written books about how you can "make your own reality" by visualizing things.

People are free to say what they like. But serious physicists and science writers tend to shy away from that kind of conclusion. We sometimes even denounce such books as "pseudoscience", often with good cause.

Yet at the same time, we don't want to move too far the other way and deny the astonishing fact that everything that seems real apparently isn't: a truth confirmed in the eyebrow-raising but highly consistent results of quantum physics experiments.

Relatively Odd

Quantum physics is not the only huge, intriguing area of modern science. There's also relativity, forever associated with Einstein, that's often described as the other big science breakthrough of the 20th century.

Relativity is certainly another concept that seems to defy logic.

It has mind-expanding implications about the nature of time, and indicates that empty space (even a vacuum) can be stretched, squashed or squeezed.

Many scientists stopped thinking of time and space as separate concepts after reading the theory of relativity.

Then, there is the issue of how time works. We used to think that a minute lasted a minute and the same was true for everyone. But Einstein taught us that that was true only sometimes, depending on facts such as where the person measuring the minute was standing, and how fast they were moving. If you and I were in different aircrafts and both of us timed a minute with atomic clocks (the most accurate type of timepiece), they may measure different periods and move out of sync — yet both would be exactly right! Logically it seems impossible. But it's a repeatable experiment that always produces the same results.

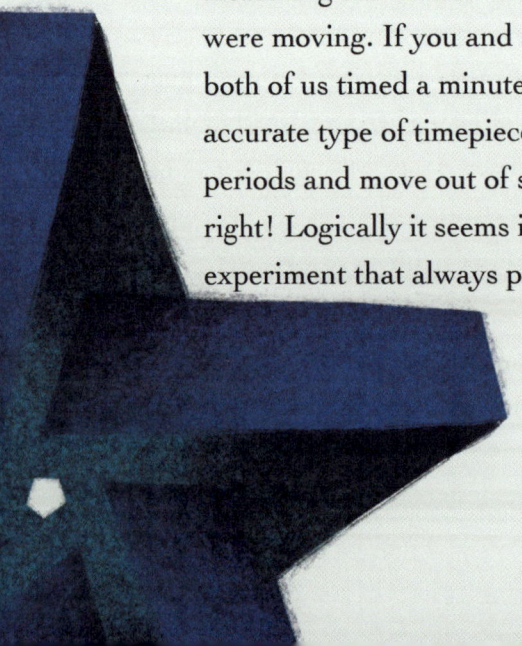

Not Too Hard

Both quantum physics and relativity are widely assumed to be too complex and esoteric for the average reader.

But I don't think so. This brief book, I hope, explains these subjects simply; telling the straightforward, true stories of their discoverers at work at each key stage. I hope it makes these topics accessible to a wider range of people, including general readers, young readers, readers for whom English is not their first language, or any non-scientist who is intrigued by how reality actually seems to work, but who shies away from physics and mathematics.

In addition to this ambition, this volume differs from others on these topics in more ways.

Science Belongs to All of Us

Most volumes on these subjects focus very narrowly on a specific series of breakthroughs in offices and laboratories in Europe in the early 1900s.

That makes sense, but those discoveries were foreshadowed and underpinned by findings that had been gathering pace for millennia, around the world.

So we have made our timeline much, much longer. Relativity, for example, is rooted in the findings of Galileo, written 400 years before Einstein's more famous writings on the same subject.

Furthermore, quantum physics deals with the nature of reality, a topic that overlaps with philosophy, and ways of thinking around the world. The early Hindu mystics, for example, were thinking and writing about it several millennia ago, as were the Chinese. And they did so in ways that were very similar to the thinking of the 20th century fathers of quantum physics.

So our geographical spread is not limited to Europe, but is more global, and our timeline is all of human history. (We like a big canvas!)

Moments of Discovery

Big breakthroughs are nearly always rooted in strings of smaller findings. At each stage of development of our present understanding of the world, we had a person looking at something and thinking:

"That's impossible — that makes no sense at all! And yet…"

And thus we get the structure for this volume: 21 Impossible Things.

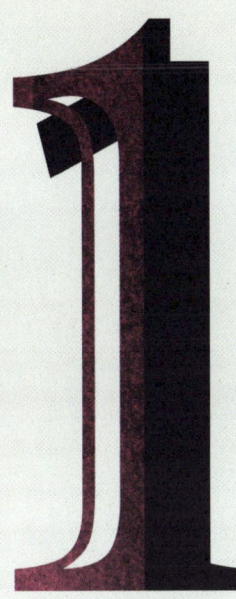

Reality is Not What it Seems to Be

Back to our two opening questions:
1) Is the world real?
2) Is one second one second long?
It seems impossible that the answers to these questions could be anything but "Yes" and "Yes".
And yet…

ONE COLD AUTUMN morning, almost 100 years ago, the world's greatest physicists converged by train on Brussels, a city in Belgium, to see if they could agree on an answer to a very strange mystery:

Is the world we live in real?

Now you might think that that's a very odd question. Of course the world is real. If it wasn't real, the physicists wouldn't be here walking around in it to ask the question, would they?

Well, members of that elite council of scientists would have agreed with you — if you had asked it just a few years earlier.

But the results of a series of experiments had left them with some very curious results. Particles did not seem to be there at all, until someone interacted with them. And since everything is made of particles, the discovery indicated that all objects, yes, every single thing in the world, were not real until people looked at it.

The experiments were repeatable, the answers were consistent — and the implications were astonishing.

You'd think that the physicists would be stunned by this, but they weren't as surprised as one might expect. Many had spiritual backgrounds, mostly Christian or Jewish, and were open to the idea that there were deep metaphysical secrets that we had not yet discovered about reality. Others were keen readers of ancient Eastern philosophy.

Now, when we talk about very clever people, many folks think of Albert Einstein – a physicist whose mustached face and messy white hair are famous around the world. He was there at the meeting we are discussing, stepping off the train at Brussels North Station with his battered suitcase.

At the meeting, Einstein (whose hair had not yet turned fully white) made it clear that he thought the world was definitely real. He came to be considered the informal leader of the "our reality is real" school of thought.

His main opponent was a long-faced, intense Danish man named Niels Bohr, who was responsible for several of the breakthrough ideas about the tiny particles that make up you, me, and everything else. Experimental results had made him very suspicious about the regular understanding of reality.

Think about that – a science argument between two people, one of whom is Albert Einstein! You'd think that Einstein's opponent would lose the battle pretty quickly.

But amazingly enough, *that didn't happen*. The group discussed the topic at length, but were completely unable to reach an agreement.

At the end of the week, they headed back to the two Brussels railway stations without answers, knowing only that reality was a puzzle, stranger than it seemed on the surface.

In the years that followed, the quiet Danish man and other scientists working with him developed a new field of science called quantum physics. It indicated that everything seemed to be made of nothing, and "matter was made out of ideas…", as one of Niels Bohr's colleagues, Werner Heisenberg, said.

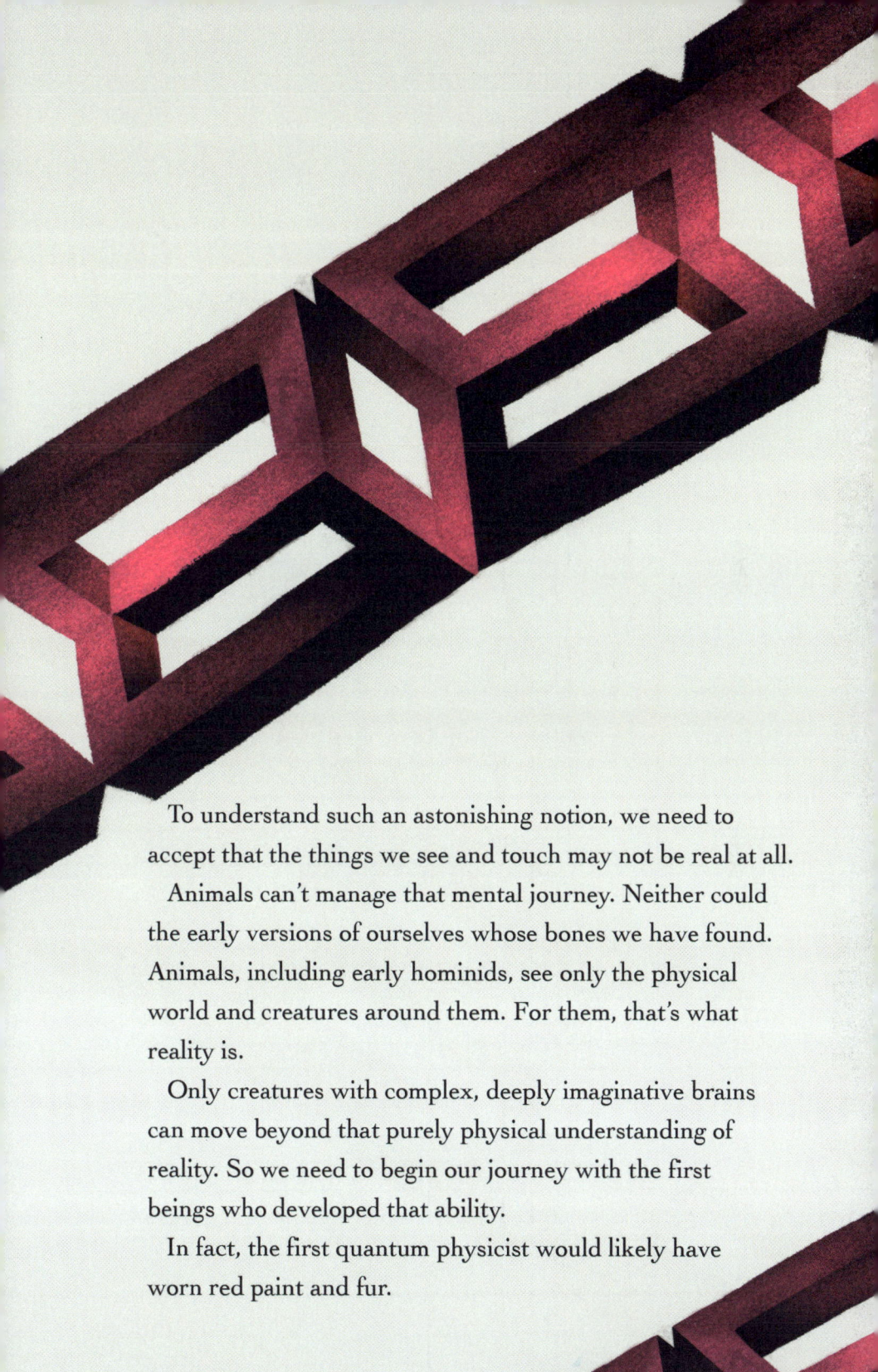

To understand such an astonishing notion, we need to accept that the things we see and touch may not be real at all.

Animals can't manage that mental journey. Neither could the early versions of ourselves whose bones we have found. Animals, including early hominids, see only the physical world and creatures around them. For them, that's what reality is.

Only creatures with complex, deeply imaginative brains can move beyond that purely physical understanding of reality. So we need to begin our journey with the first beings who developed that ability.

In fact, the first quantum physicist would likely have worn red paint and fur.

2
Seeing What Isn't There

The majority of animals see what's there.
But one creature sees what isn't there.
Logic tells us that this should be a massive
disadvantage. And yet this turns out to be
our most important skill.

A LONG TIME AGO, there likely existed a pair of children who made a much wider range of noises than their peers. They started with a few sounds as babies, but developed a wide, complex repertoire of sounds as they grew into young adults.

Other members of the tribe were fascinated by these youngsters because the noises they made to each other seemed to give them superpowers! Seriously.

Now this would have happened at least 50,000 years ago, and perhaps several times over before it began to spread widely.

In those days, animals communicated the same way animals communicate now — with warning calls (such as roaring or barking), pay-attention-to-me noises (such as mewing) and quiet hostile noises (such as growling).

But a branch of humanity probably quite suddenly developed a sound system which was on a totally different level of complexity, what we now call language. They were able to create very detailed, lengthy, complicated pieces of information, and exchange them; quickly and efficiently planting these complex pieces of information in each other's brains.

Where did it come from? No one knows for sure, but some scientists think that a change in the brain, perhaps a mutation, a technique or habit, triggered a new way of thinking.

Scientists call it "symbolic thinking", since a vocalized sound (or a scribble or a gesture) made by one person could represent an object or concept in the mind of another person. This was clearly related to the development of a powerful form of imagination.

These "new" people had heads full of rich, complex ideas, and we don't know exactly where or when they first emerged, but we know it started making a big difference in human society by about 40,000 years ago, so it happened before that — perhaps 50,000 years ago.

Specialists say it would probably have been the case that two of these "new" people would eventually have been born into one family (perhaps as twins), or one clan (perhaps as cousins). The pair would naturally have developed a system to exchange these high level ideas.

And that would have been the moment that language was born.

We can imagine a pair of bright young women, perhaps, pondering ideas and chattering to each other in a language they made up together, while other early humans stared at them, puzzled.

Their inner language ability (handling concepts) and their outer language ability (communicating them to each other) would have been like a superpower.

Here's how it would have worked.

Up until that time, if a member of the group found something useful for the clan, such as an unoccupied grove, oasis or cave they could live in, or a good food source, they would have to take their relatives along and show them physically.

But not our twins. The possession of language made a huge difference.

One could simply say to the other: "Sister, this morning I went for a walk directly towards the sunrise and found a beehive full of fresh honey by the third pine tree on flat-top hill. There's a cave on the right, obscured by a red-fruit bush, that we could make use of. Why don't you take the others there and show them, while I prepare a meal?"

"Okay," the other could reply. "I saw a small flock of jungle fowl nesting 200 paces beyond the river bend, near the second apple tree. You can catch a few of those for cooking."

It was this ability to instantly transfer complicated ideas about things far away that would have seemed like magic.

The ability somehow spread — we don't know exactly how — and humans quickly developed music and art and spirituality and science and so many wonderful things in quick succession. The result is the modern world we see today.

But the most important part of this change, for the purposes of our story, was this: humans accepted that there was much more to reality than things that could be seen and touched. They mentally moved away from the physical world and started to "live in their heads".

The complexity of their inner worlds began to match the complexity of the outer world.

At first, the ideas may have been purely practical, but (humans being who they are) they would soon be thinking and philosophizing about life and the nature of reality, just as we do today.

The new way of thinking changed everything.

When a creature dies, its fellow creatures see a dead body. But these "new" humans saw a puzzle. They saw a physical body that appeared to have been emptied of its soul. Where had it gone?

3

A World of Dust

People had no doubt that the world was made of solid things such as earth and stone and wood and other materials. But two men realized this only *seemed* to be the case. Reality was very different.

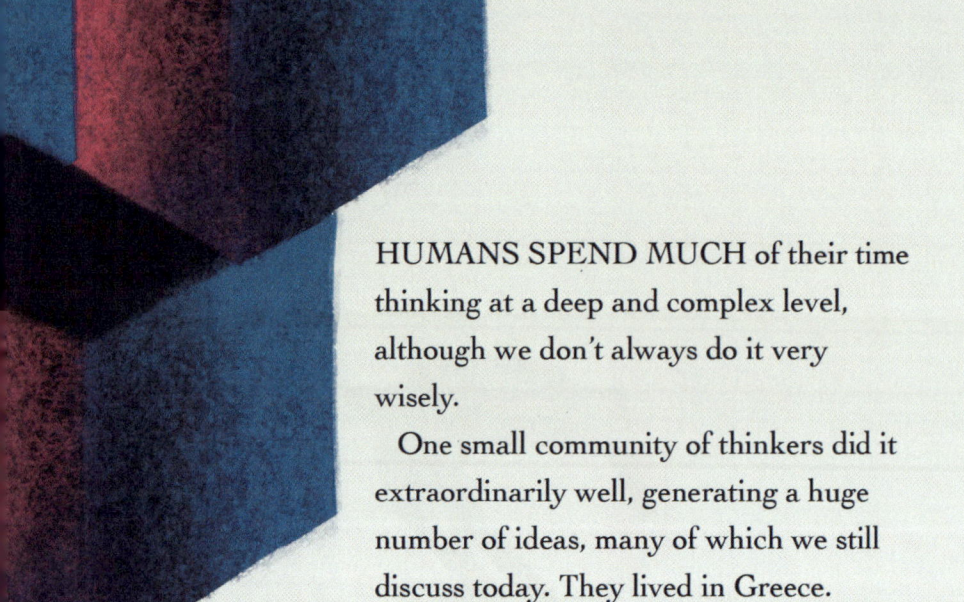

HUMANS SPEND MUCH of their time thinking at a deep and complex level, although we don't always do it very wisely.

One small community of thinkers did it extraordinarily well, generating a huge number of ideas, many of which we still discuss today. They lived in Greece.

One day, about 2,500 years ago, a man named Democritus was busy telling people that the world was not made of the stuff it appeared to be made of: earth, wood, metal, stone, and so on.

Democritus was who we might think of today as a rich hippie backpacker (he inherited wealth and spent much of it travelling to other countries). Yet he was also a good student, working hard to refine ideas that he and his teacher Leucippus had developed.

They had worked out that everything that existed could be crumbled down and down and down to smaller pieces.

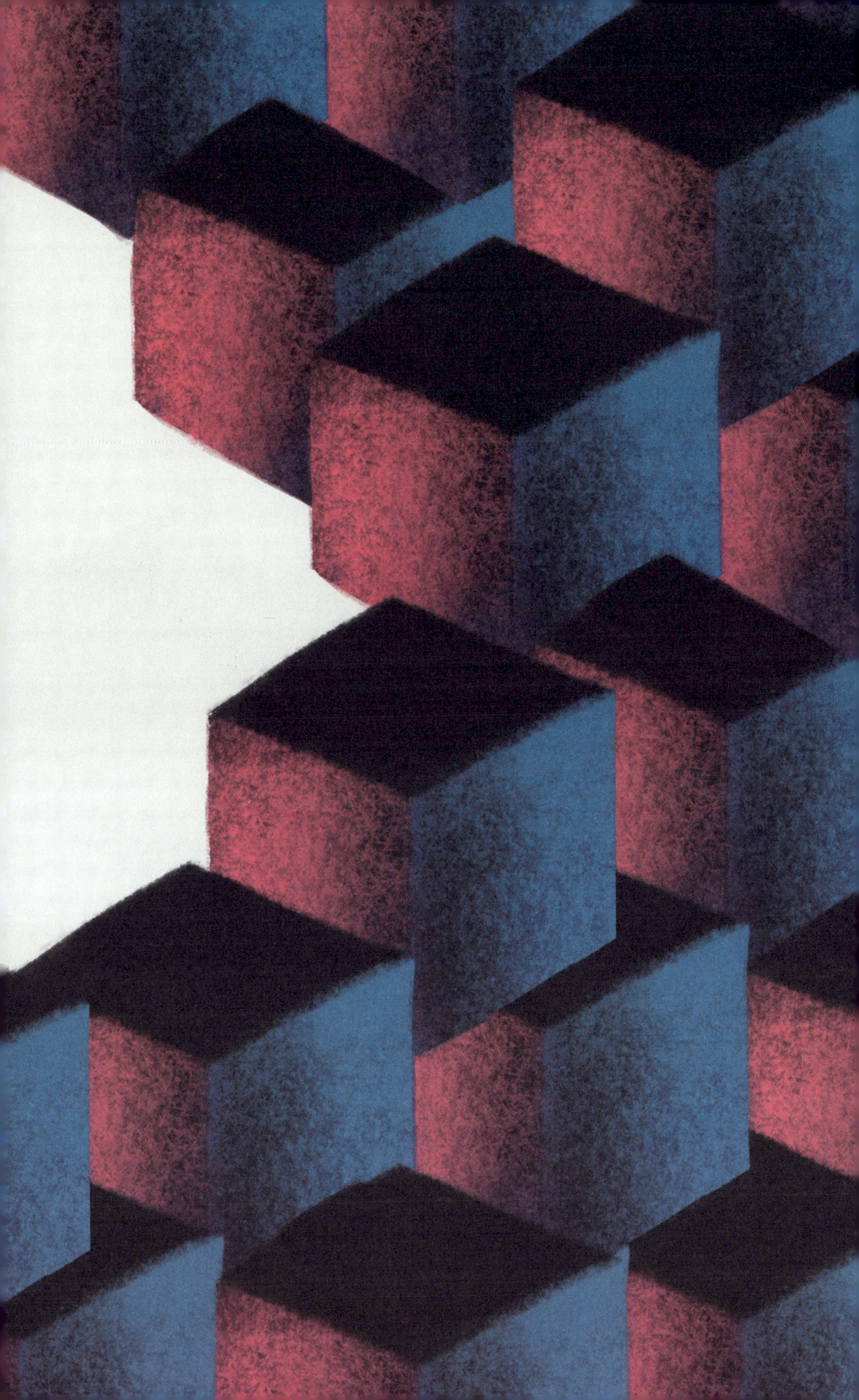

20   21 Impossible Things

You could logically keep crumbling until you ended up with invisibly tiny particles which could not be cut further. Logically these indestructible particles of dust must be the building blocks of reality.

Really? But that makes no sense, people complained. How could things as different as iron and water, for example, be made up of the same tiny bricks? They're so different.

Democritus suggested that there were different types of particles. The ones that make up metal were tightly hooked together while the bits that made water were slippery and easy to split.

But he had a question for himself: What should we call these bits?

For now, let's leave him pondering, and move to the other side of town, where two other brilliant men, also a teacher and a student, were developing ideas about the nature of reality.

Plato said that people thought they knew what reality was, but all they were really doing was looking at a shadow of what really exists, which was very different, and far, far more complex.

His student Aristotle was having equally revolutionary thoughts, but about the nature of time. He had worked out that everything that happens is caused by something. Therefore reality can be seen as a long chain of cause-effect, cause-effect, cause-effect, stretching back to the beginning of time.

So logically, there must have been a first thing, right?

And that would have to be an Uncaused Cause. The idea intrigued him. Who was this and what did they actually do to get the whole system rolling?

Aristotle's amazing ability to make intelligent conjectures about the world around him made him famous, and he is still seen today as one of the greatest scientific thinkers in history.

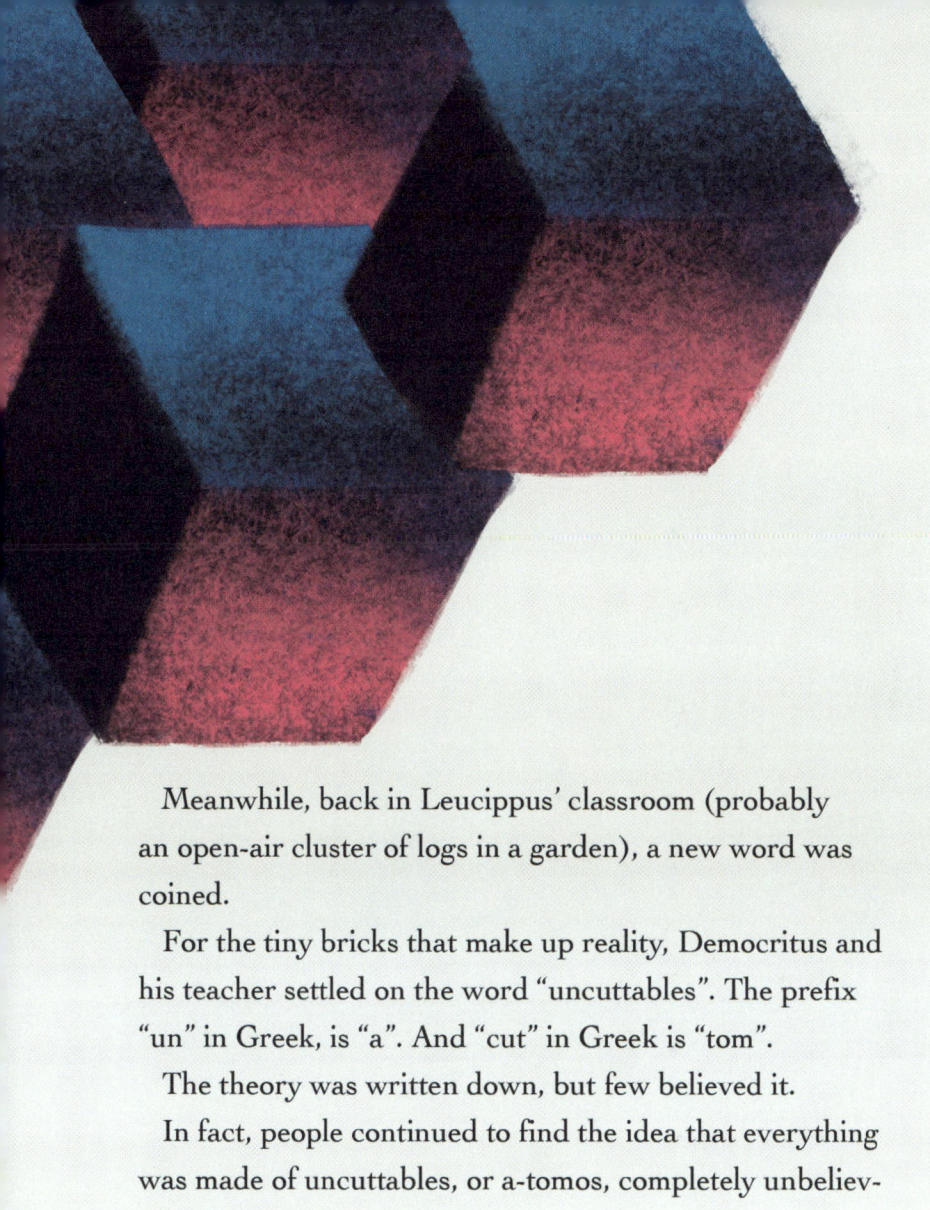

Meanwhile, back in Leucippus' classroom (probably an open-air cluster of logs in a garden), a new word was coined.

For the tiny bricks that make up reality, Democritus and his teacher settled on the word "uncuttables". The prefix "un" in Greek, is "a". And "cut" in Greek is "tom".

The theory was written down, but few believed it.

In fact, people continued to find the idea that everything was made of uncuttables, or a-tomos, completely unbelievable for more than 2,000 years.

It wasn't until the modern era that people realized that the world, quite possibly, really was made of uncuttables: items known in English, from the Greek, as *atoms*.

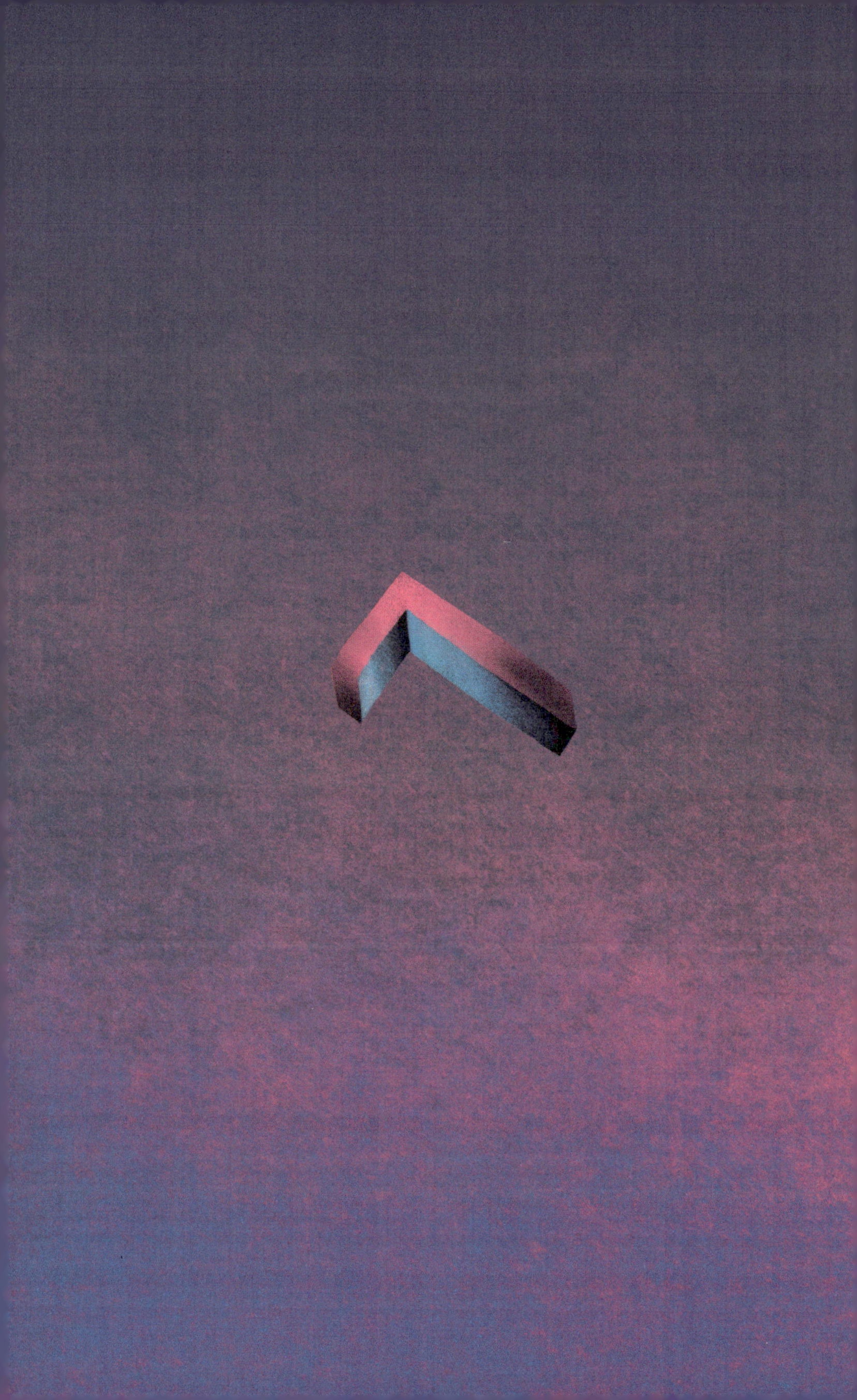

4

Reality is a Complicated Place

Logically, there's no such thing as free will, and stuff is made out of stuff. But early philosophers realized these self-evident truths were probably not true at all.

"WHAT SHALL I HAVE for breakfast? And what is the nature of reality?" The Chinese philosophers of ancient days quickly moved from the smallest, most practical questions, to the largest, broadest ones — often tackling both at once.

Let's picture Mozi, who lived more than 2,400 years ago, taking his morning tea and going to sit in a Chinese garden; he and his peers felt there was something special about gardens, an idea many other cultures seemed to develop simultaneously.

Mozi was wrestling with a specific problem. The universe seemed to be made of stuff, and all stuff seemed to be subject to the laws of cause and effect. Everything was caused by whatever came before it.

That meant the universe worked like a big machine, and we were just cogs of the machine. But if that were true, free will could not exist.

In China, like other lands, people at both extremes of wealth often made an argument which went like this: Kings ruled, so kings said that the law of cause and effect must mean they were destined to rule. Similarly, peasants struggled, so peasants said (probably with less enthusiasm) that they were fated to struggle.

Mozi became convinced that *Tian* (a word he used to mean both Heaven as a place and Heaven as a metaphor for God) was actively good, and humans always felt they should strive to better themselves. Yet they could not make good choices without free will.

So this seemed like a mechanical universe in which every event followed inevitably from every earlier event. But it wasn't: it was a flexible one in which individual decisions mattered. Reality was about choices.

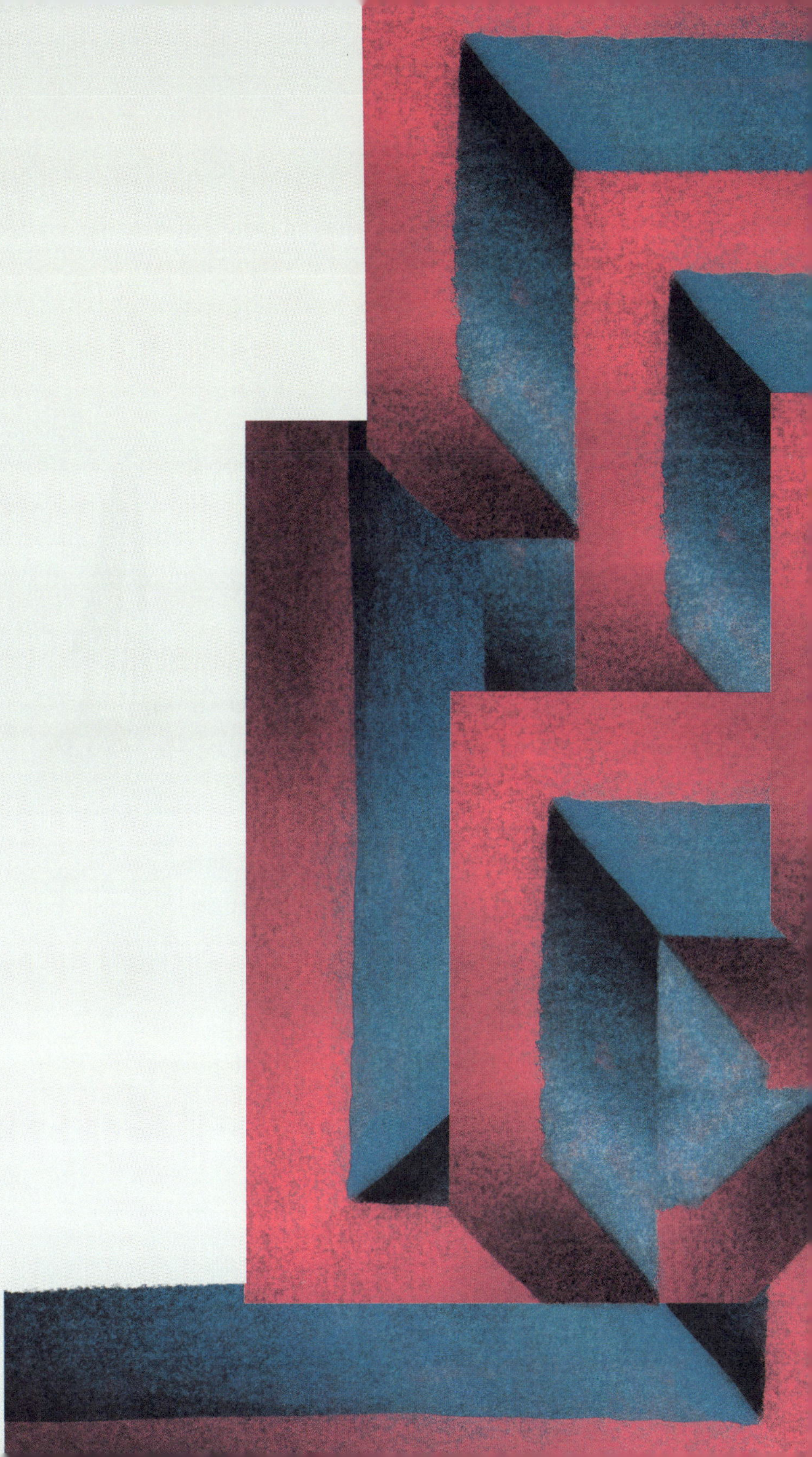

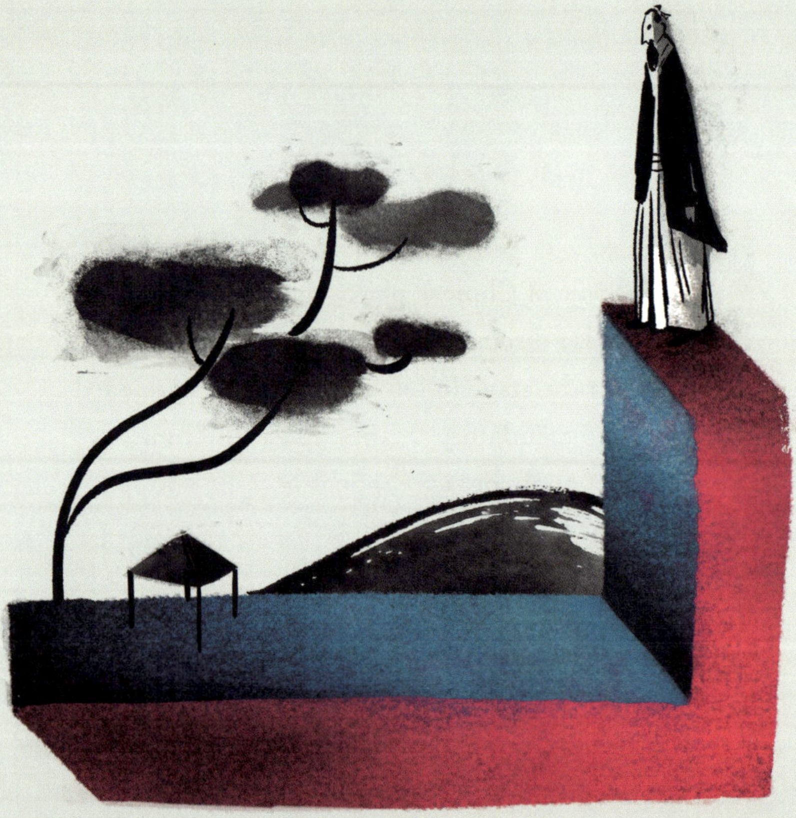

Another pair of Chinese philosophers, Laozi and Zhuangzi, also discussed the nature of reality. They realized that the stage in which everything happened involved two mandatory factors: space and time. Everything you do or say happens at a specific spot at a specific time. No exceptions.

But they noticed that humans everywhere developed belief systems, implying that there was another layer of reality. This faith was so universal among the planet's most intelligent creatures that it must have value and was probably true.

To the Chinese philosophers, this other layer was The *Dao* (The Way), a place and a process which was not limited by space or time. It was everywhere and nowhere, every-when and no-when.

From the *dao* came *qi* which is "the energy that comprises all things". So although objects appear to be made from matter, they are actually all made from energy. *Dao* rhymes with "wow" and *qi* is pronounced "chee".

But they did not know the exact recipe that transforms one into another (we had to wait for Einstein and others to work that one out).

When we read the writings of these philosophers of ancient times, we can't help but be amazed at just how long ago people were writing texts, identifying the key questions about reality that we typically think of as modern puzzles.

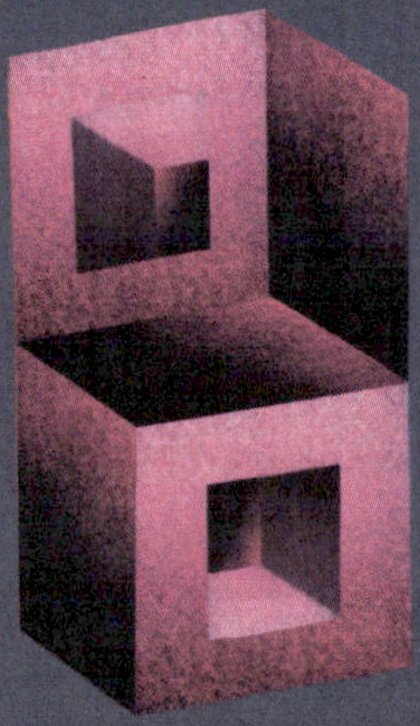

5

Stepping Beyond Reality

Dreams regularly fool us into believing that we are moving around in a 3D space filled with real objects. It would be impossible to do that trick on a life-long scale. Right?

ONE NIGHT, A VERY long time ago, a man in Asia had a lucid dream.

His dream world seemed so real. He could see every blade of grass. He could see the texture of the skin on his hands. He could feel the breeze on his face, and smell the scent of flowers.

There were no detectable differences between the real world and the dream world.

When he woke up, he realized there was a puzzle here.

Have I woken up?

Perhaps I have not. Maybe it is only the dream version of me which has woken up?

The interesting thing is this: with dreams which are detailed enough, there's no way you can tell.

Who was he? We can see from ancient writings that several people had this experience and wrote about it, particularly in India and China.

In China, it was philosopher Zhuangzi, who we mentioned in the previous story. He woke from sleep and realized something over his morning tea.

A person can have a dream of himself waking up and analysing the dream he has just awoken from—but all the while, he is still asleep, but just in another layer of dream.

Then he realized that he might be describing himself at that moment.

Perhaps we only think we are awake in a physical world, Zhuangzi decided.

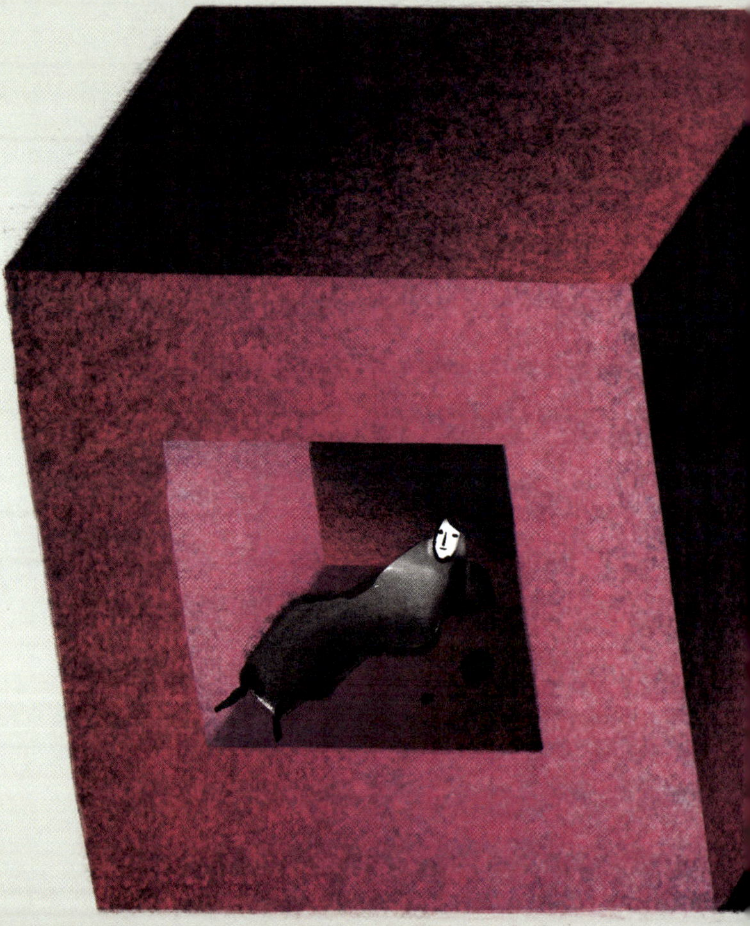

Stepping Beyond Reality   37

A more likely truth is that none of us ever are! Reality itself can be seen as another layer of dreaming.

In India, a Buddhist named Vasubandhu had a thought along the same lines, and said that physical stuff likely didn't exist at all.

Another Hindu writer ran with the notion. Reality itself is definitely a type of dream, the wisest of us agree—but since we all share it, whose dream is it?

The universe itself must be dreaming it, he realized.

That would make reality a dream in the mind of God. He called the creator god Brahman.

To the average person, these ideas may seem to be rather fanciful and not very scientific. Surely, it makes no sense to think that reality is just another dream?

Yet, these ideas were valuable—and would prove to be deeply scientific in the most fundamental way: moving humanity away from the possibly false notion that reality consists of physical beings moving around in a physical world.

These notions about reality were widely adopted to form cultural frameworks, which some call faiths or religions. They were key to developing humanity's understanding of how a being with a complex consciousness should live.

Human children, like animals, see reality as a simple physical place, and think we live in a world of straightforward, what-you-see-is-what-you-get truths. But as we grow and acquire wisdom, we realize that we don't live in a world of facts at all. We live in a world of intuition, feelings, emotions, mysteries, connections and perceptions we never feel we fully understand.

But how far should we distrust physicality?

Could Vasubandhu be right? Did any physical object really exist at all?

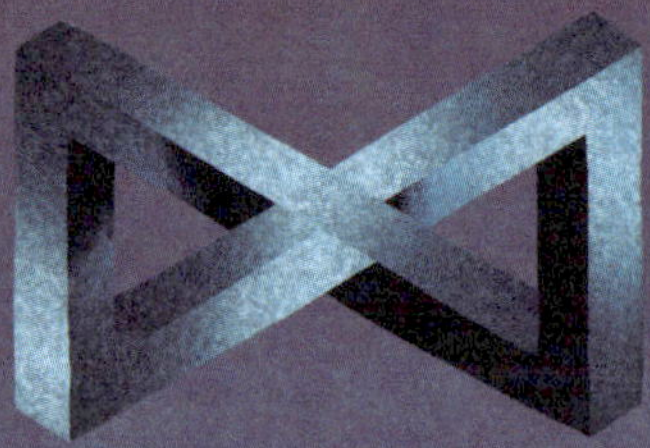

6

An Invisible Dimension Exists

Even today, a lot of people find it impossible to believe that there is another dimension of reality that we cannot see. But many early thinkers realized the truth.

THE GREEK PHILOSOPHER Plato, teacher of Aristotle, believed that there were secret halls buried far below a temple in Egypt.

They contained documents up to 9,000 years old from the people he thought of as ancient thinkers —individuals such a legendary wise man named Hermes Trismegistus.

Those chambers have never been found, but some of the things Plato's mysterious thinker wrote did survive, dug up in other places.

They show that Hermes Trismegistus realized that all communities develop ways of understanding reality.

These cultural codes shared a central idea, Hermes realized. This was it: There are TWO realities, the one in which we live and a different one. For example, the Chinese talked of Heaven (Tian), the realm of the supreme deity Shangdi. The Jewish people talked of Sheol, the place of the dead. Hindus talked of the higher planes.

(Christianity and Islam had not been developed at the time.)

In every case, our physical reality felt real, but the other one was more important—and decisions made in either reality resonated in the other.

This could be summarised in just four words: "As above, so below." Trismegistus, or others working on the same theory, later added another phrase: "As inside, so outside."

Different groups of people said these phrases in various ways, but all meant to say the same thing. For example, the phrase "On earth as it is in heaven", was found in the writings of several ancient Jewish communities.

The phrase "As inside, so outside" suggested that the thoughts and decisions you make somehow have an effect on outside reality.

You can see that these ideas are similar to the notions mentioned earlier in this book. "As above, so below" foreshadows the modern idea that that this layer of reality is simply a projection from a different, invisible one.

"As inside, so outside" is intriguingly similar to the modern discovery that measuring a particle makes it solid, implying a connection between our inner decision to do something and outer reality.

It's interesting to note that many different cultural groups claim Hermes Trismegistus to be an ahead-of-his-time presenter of their own ideas, sometimes under different names, such as Thoth (Egyptians) or Idris (Muslims) or under his full name (Christians).

An Invisible Dimension Exists 45

"Trismegistus" is Latin for "Three Times Great".

Why did he have such an odd name? Archeologists looking at an old temple on the banks of the river Nile in Egypt may have found an answer. Some of the carved writing, when translated, turned out to be a name and a title: "Thoth The Great The Great The Great".

Trismegistus, or Three Times Great, was just a shorter way to say that long surname. Clearly Thoth/Hermes The Great The Great The Great had enthusiastic fans.

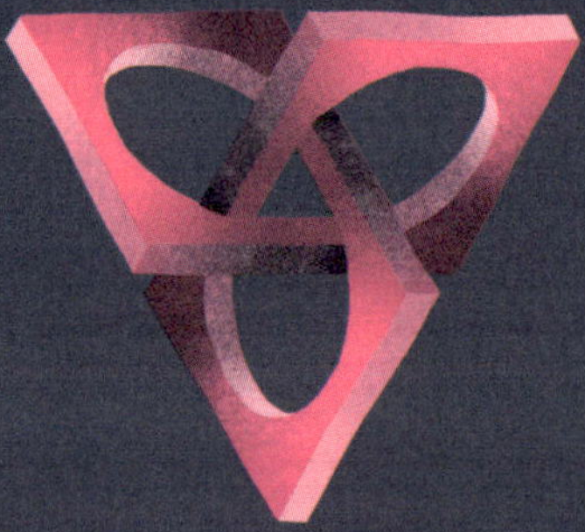

7
A Football in Space

Any creature that can push a ball soon figures out how motion works, and that's true whether you are a toddler or a baby elephant. Yet, one thinker found out that every human being's instinctive belief about this was wrong.

HERE ARE THREE things which seem obviously true:

1) An object which is not moving is an object which is still.

2) An object which is not moving won't move until something else causes it to move.

3) An object you push or kick gently might move a few centimeters, while an object you strike harder travels a longer distance—but both will stop moving eventually.

These laws of nature seem self-evident, and who could disagree?

No one.

At least, not until the 1600s, when an Italian guitar-playing wannabe priest called Galileo Galilei came along. He became a thinker and writer, and eventually shared his discovery that these indisputably "true" ideas were actually completely wrong.

Galileo had noticed something important.

The three "self-evident" points were only true where friction exists. If you roll a ball on a road with a rough surface, it will quickly stop moving.

The same ball will go further on a smooth surface, and a longer way on a pool of oil. The deciding factor was friction.

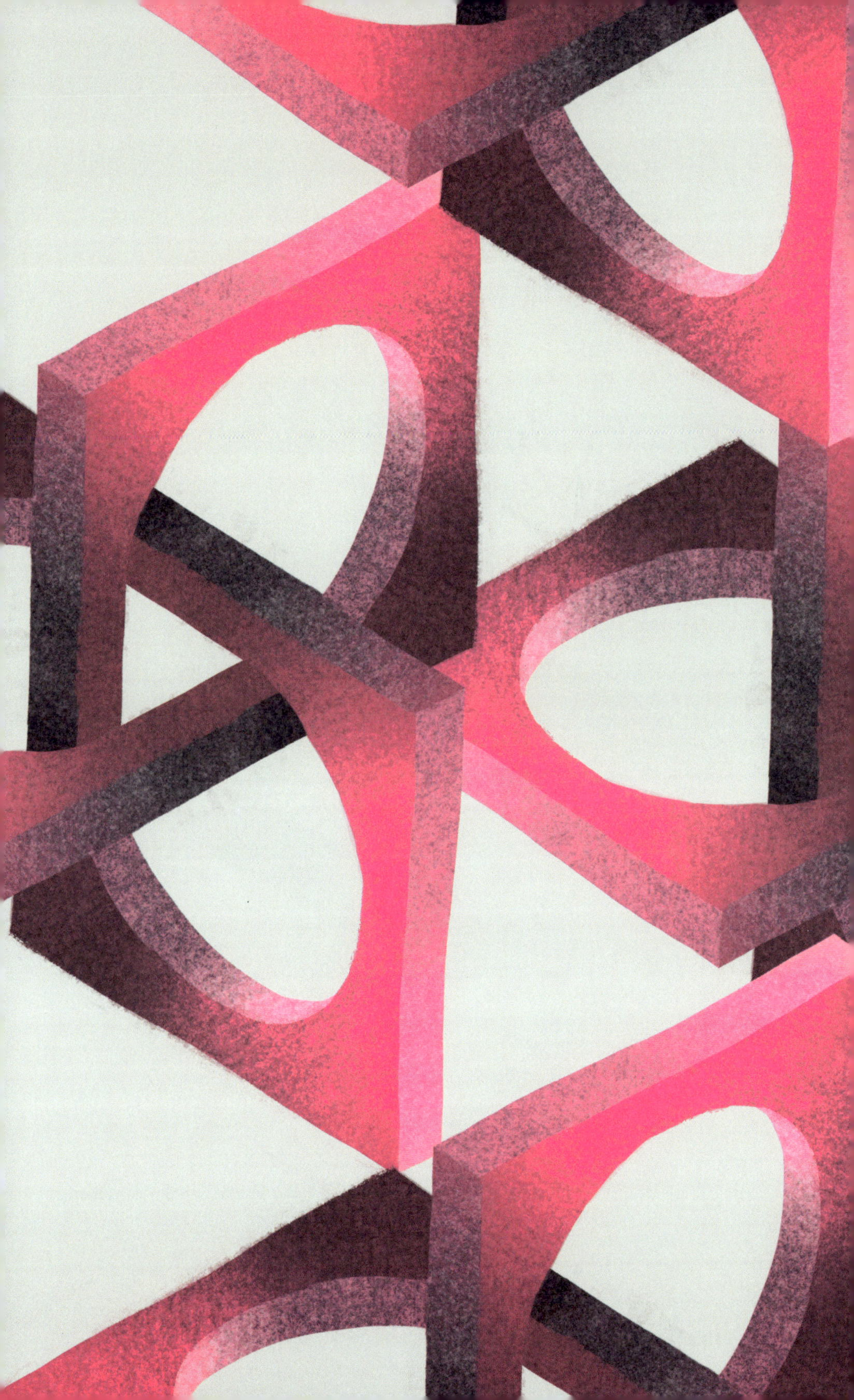

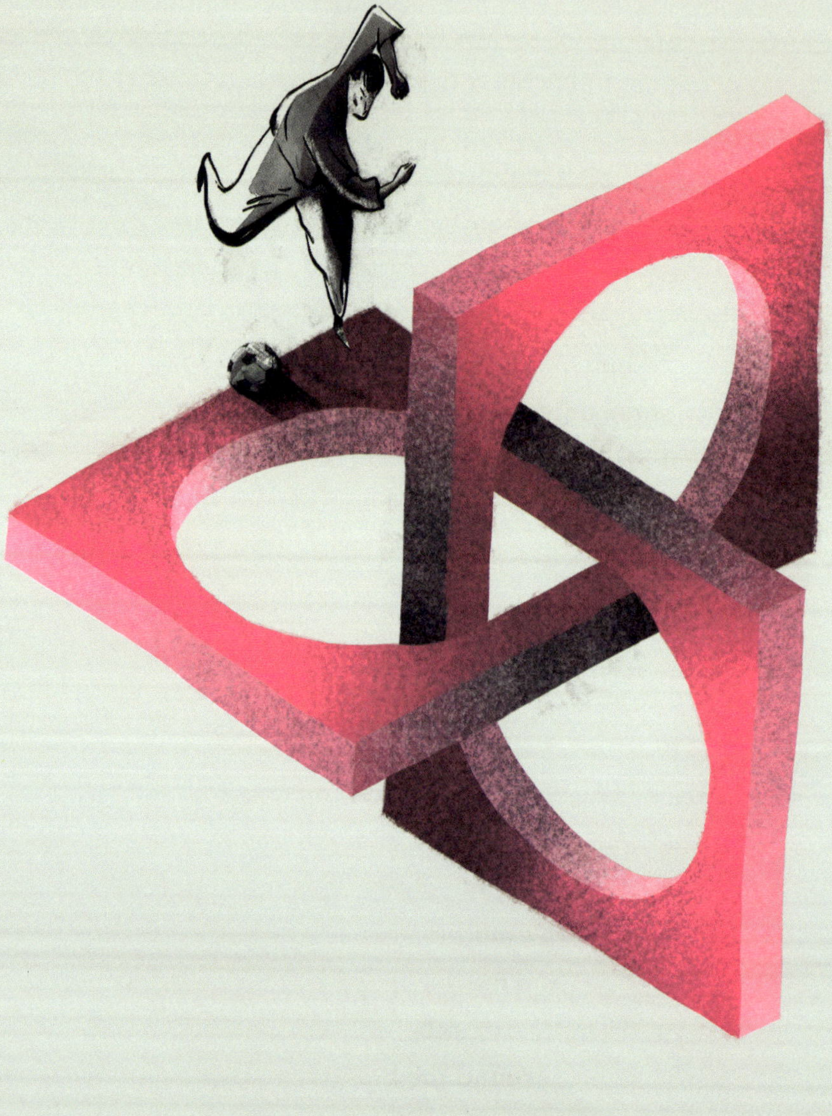

To truly understand base reality, you need to isolate things—think of each of them separately.

What if the road disappeared? What if you kicked a ball in a place where there was no friction?

Would the ball ever stop?

Actually, no! The way reality was constructed would mean that it would go on for ever and ever: a football, crossing the universe, travelling onward for eternity, he realized.

That's how our reality really worked—and our basic assumptions about motion were not true. The dimension in which God had placed us was more curious than we realized, Galileo contemplated.

That led him to another thought.

What if the universe disappeared? The ball would still be moving, but you wouldn't be able to tell it was moving—because there would be no background to measure the movement against. It would seem to be perfectly still.

The idea of an object "moving" only makes sense if there are other objects to measure it by, he realized. Everything is relative.

This idea was the birth of the set of notions that came to be known as relativity.

At the time, astronomers were debating with each other about Earth's motion.

Some said the sun moved around the Earth, as Greek philosopher Aristotle believed, and some said the Earth moved around the Sun, as claimed by a Christian astronomer named Tycho Brae.

The Pope at the time thought that Aristotle's theory was closer to the ancient Jewish view in the early parts of the Hebrew scriptures—a view shared by virtually all cultures and all religious groups.

But most Christians living in Italy, including Galileo, had started leaning towards Tycho Brae's view.

Galileo realized that the Earth could be that football flying through space.

The Earth seems to be still to one observer (such as us sitting on it), but it could be zooming along at high speed (to someone in space, such as God).

Today, we know that Galileo was right: the whole concept of stillness is impossible; it's an illusion. The Earth is moving at 107,000 kilometers an hour around the Sun. And the whole solar system itself is moving, too.

Stillness can only be achieved in our heads.

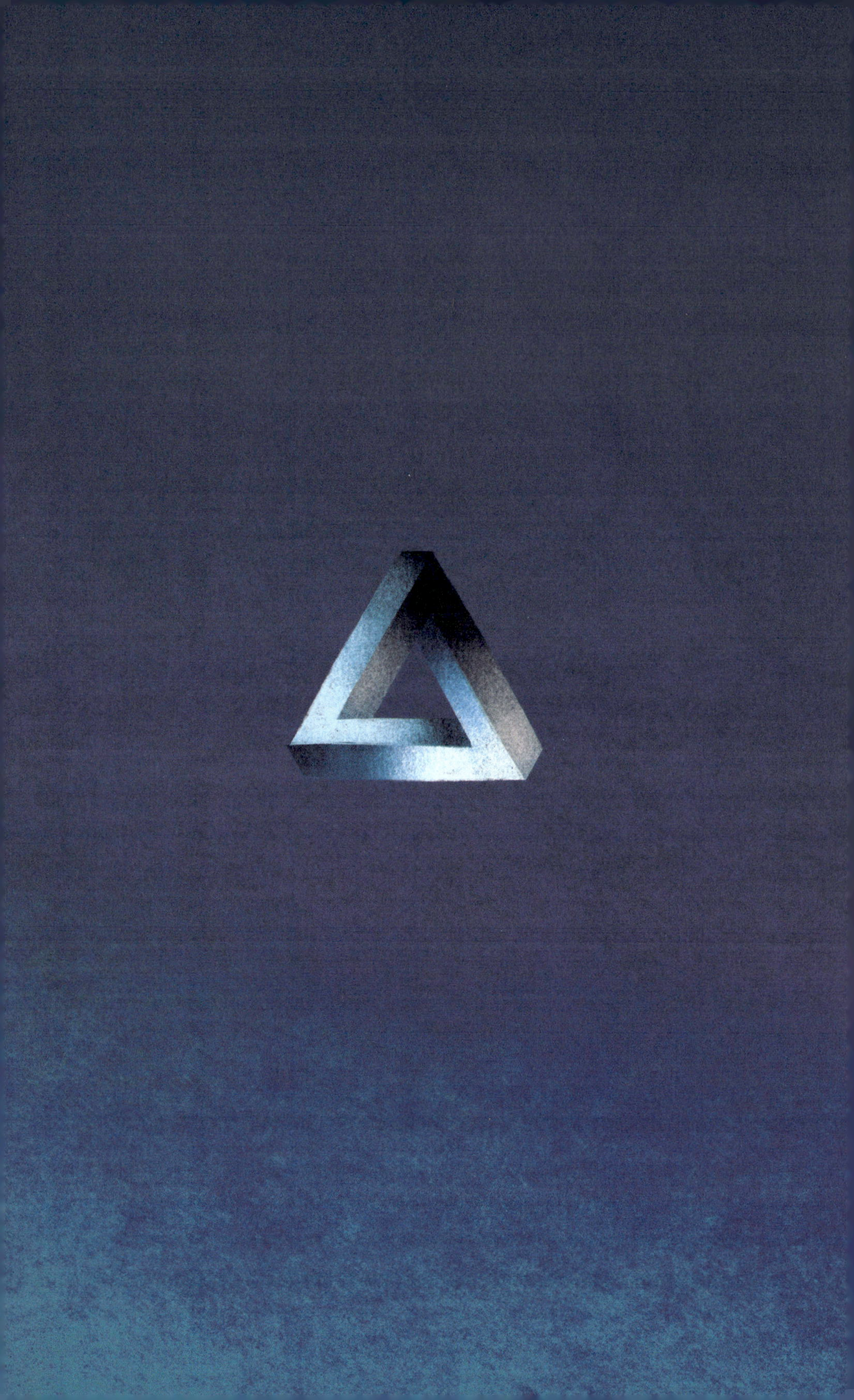

8

The Mysteriously Heavy Feather

Logic tells us that heavy things will always
fall faster than light things.
But the truth is quite different.

LONG AGO, ARISTOTLE concluded that things you dropped fell towards the center of the earth — either very quickly, if they were heavy objects (think of a rock or a hammer), or less quickly if they were light things (think of a marble or a feather).

These theories seemed so obviously "true" that they were rarely questioned.

But Galileo, the trainee priest-guitarist from the previous chapter, was a voracious reader. After reading Aristotle's works, he continued to investigate the theory. He learned that several people had noticed problems in these "obviously true" beliefs.

A Dutchman named Simon Stevin had dropped a heavy and a light object off the top of church tower, and they seemed to have landed at the same time.

Could Aristotle, one of the most respected fathers of science, have been wrong about this, too?

Simon Stevin's experiment must have been misleading, Aristotle's defenders said. The force that pulled things down at different speeds existed, but was so strong that the difference was too small to measure, they said.

Some people said that Galileo repeated the same experiment from the top of the Leaning Tower of Pisa. But his biographers suggest that he did something much cleverer.

He realized that there were shortcomings in Stevin's experiment.

60   21 Impossible Things

A feather falls slowly because of air resistance, a factor that Aristotle's analysis failed to mention.

Furthermore, if Stevin dropped the objects by hand, human error could cause a slight difference in release times.

And if the difference really was too small to measure, an experimenter would have to find a way to slow down the process, so that you could see whether there was a difference or not.

Galileo realized you could solve all three problems by making a long, smooth slope indoors. Winds don't blow inside homes, so air resistance wouldn't be a problem. A starting gate could be built. You could roll a big heavy ball and a much smaller one (think of a marble) slowly down the slope.

He tried it, and found that they both reached the bottom at the exact same instant.

Aristotle was definitely wrong. Our rational, logical assumption was wrong.

Galileo realized that our present base reality was constructed such that a hammer and a feather fall at the same speed, even though that idea seemed to make no sense.

Now, let us fast forward to 1971. An astronaut called David Scott was walking on the moon. He had brought with him a geological hammer (weighing 1.32 kg) and a feather (weighing 0.03 kg).

Since there is no air resistance on the moon, he could do the same experiment without worrying about winds at all. He told the millions of people watching on TV that he was doing this in honor of Galileo.

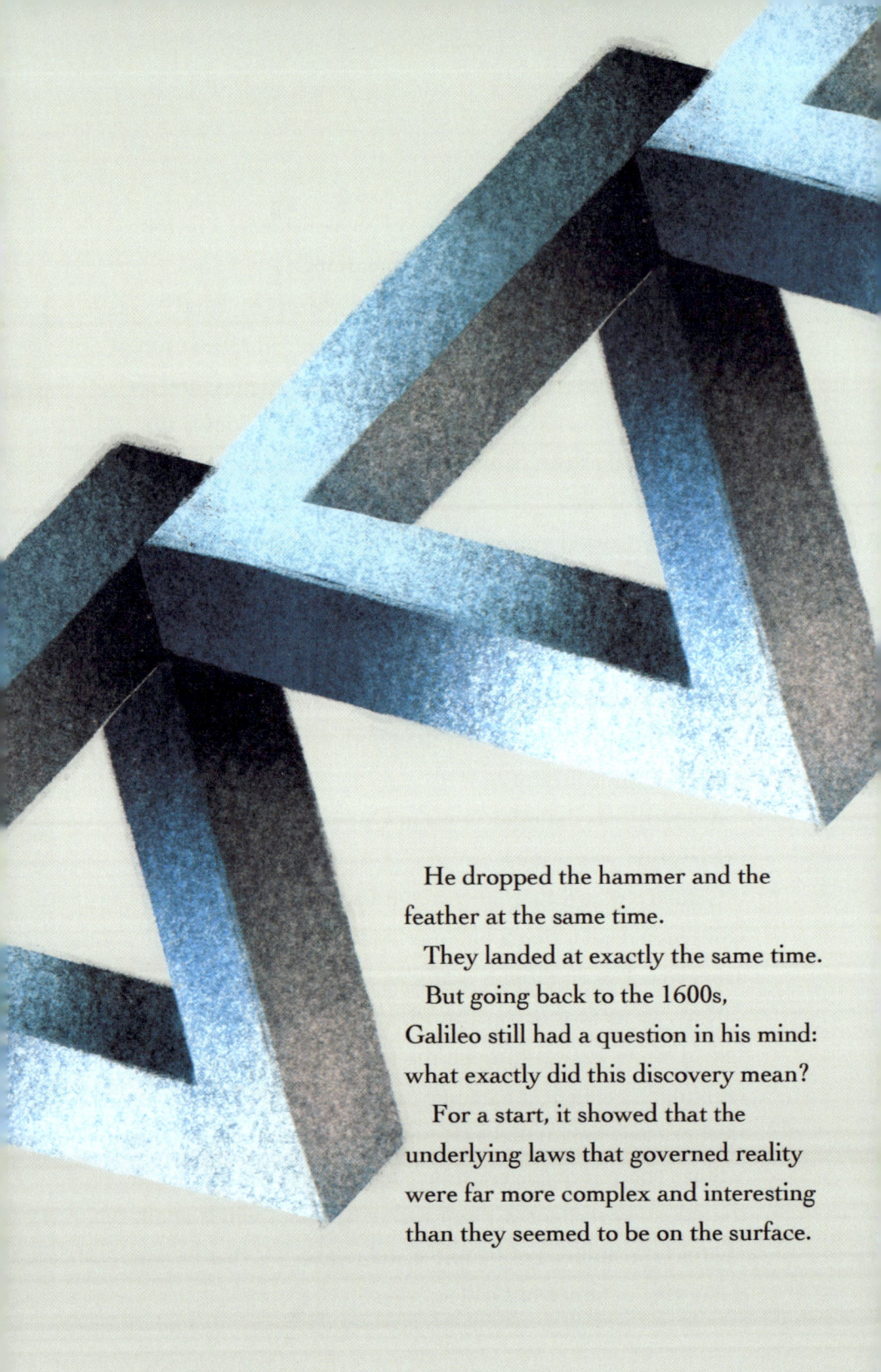

He dropped the hammer and the feather at the same time.

They landed at exactly the same time.

But going back to the 1600s, Galileo still had a question in his mind: what exactly did this discovery mean?

For a start, it showed that the underlying laws that governed reality were far more complex and interesting than they seemed to be on the surface.

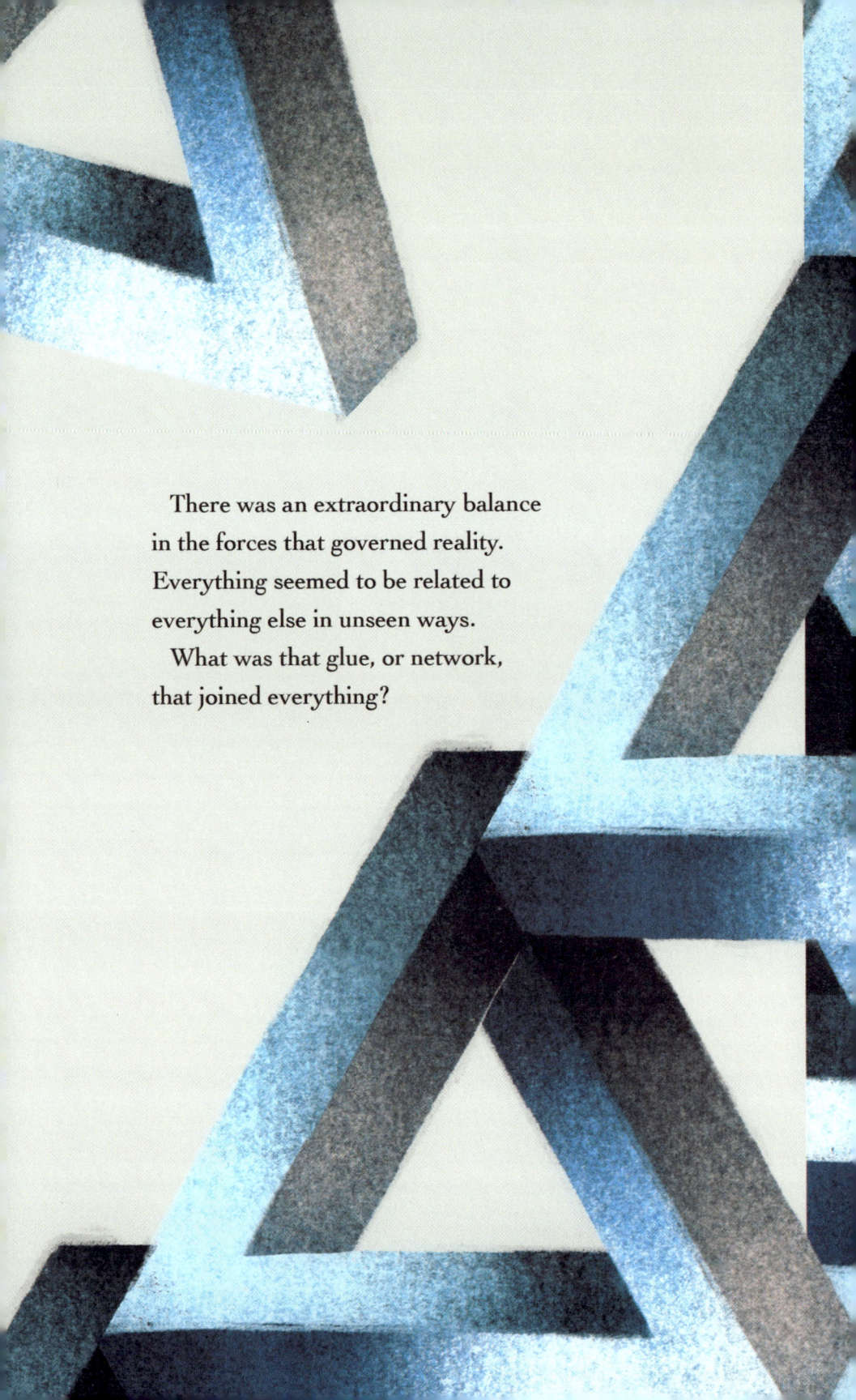

There was an extraordinary balance in the forces that governed reality. Everything seemed to be related to everything else in unseen ways.

What was that glue, or network, that joined everything?

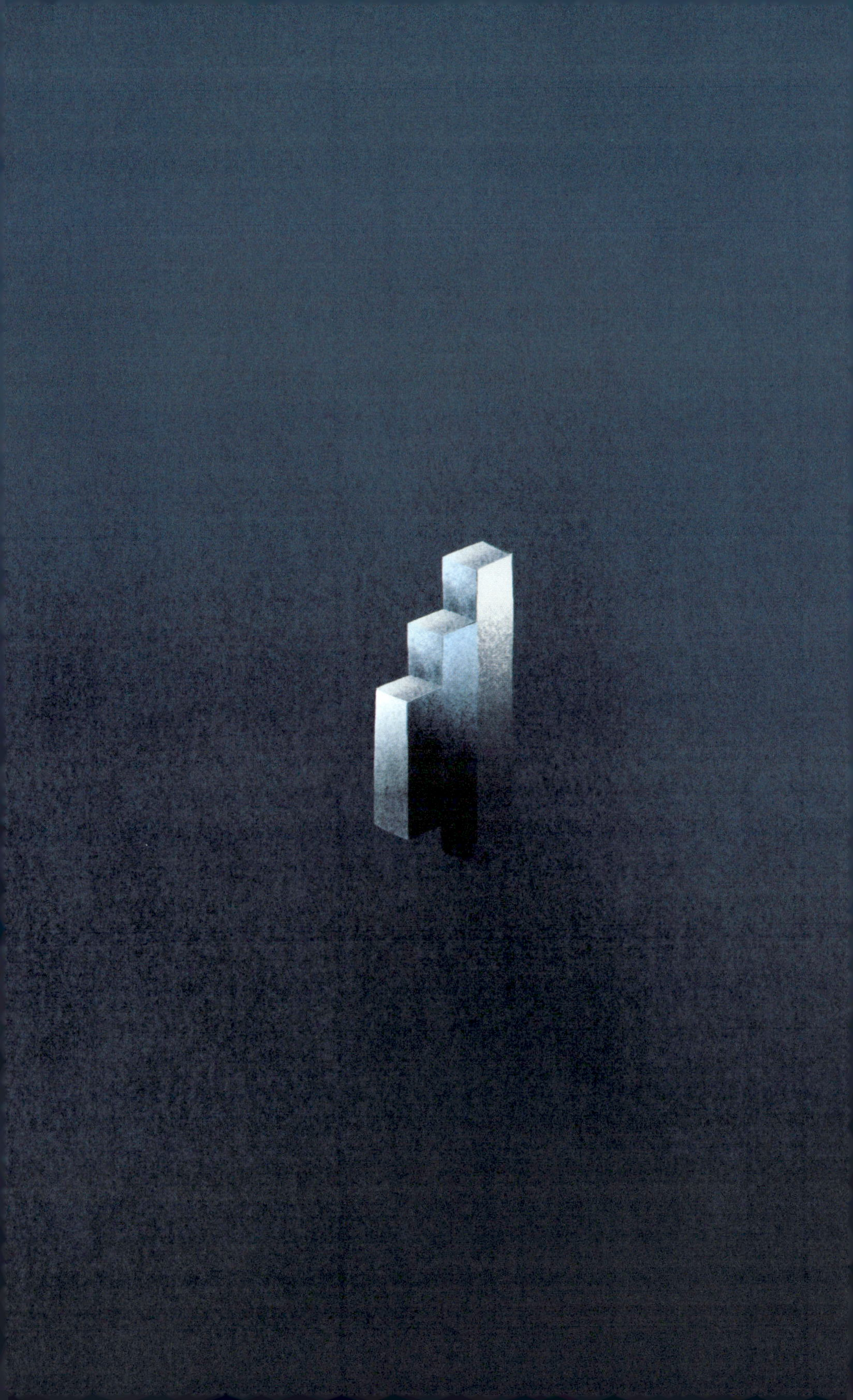

9

Keeping Body and Soul Together

Some people say you are a body with a soul.
Others say you are a soul with a body.
Both are wrong, one man said.

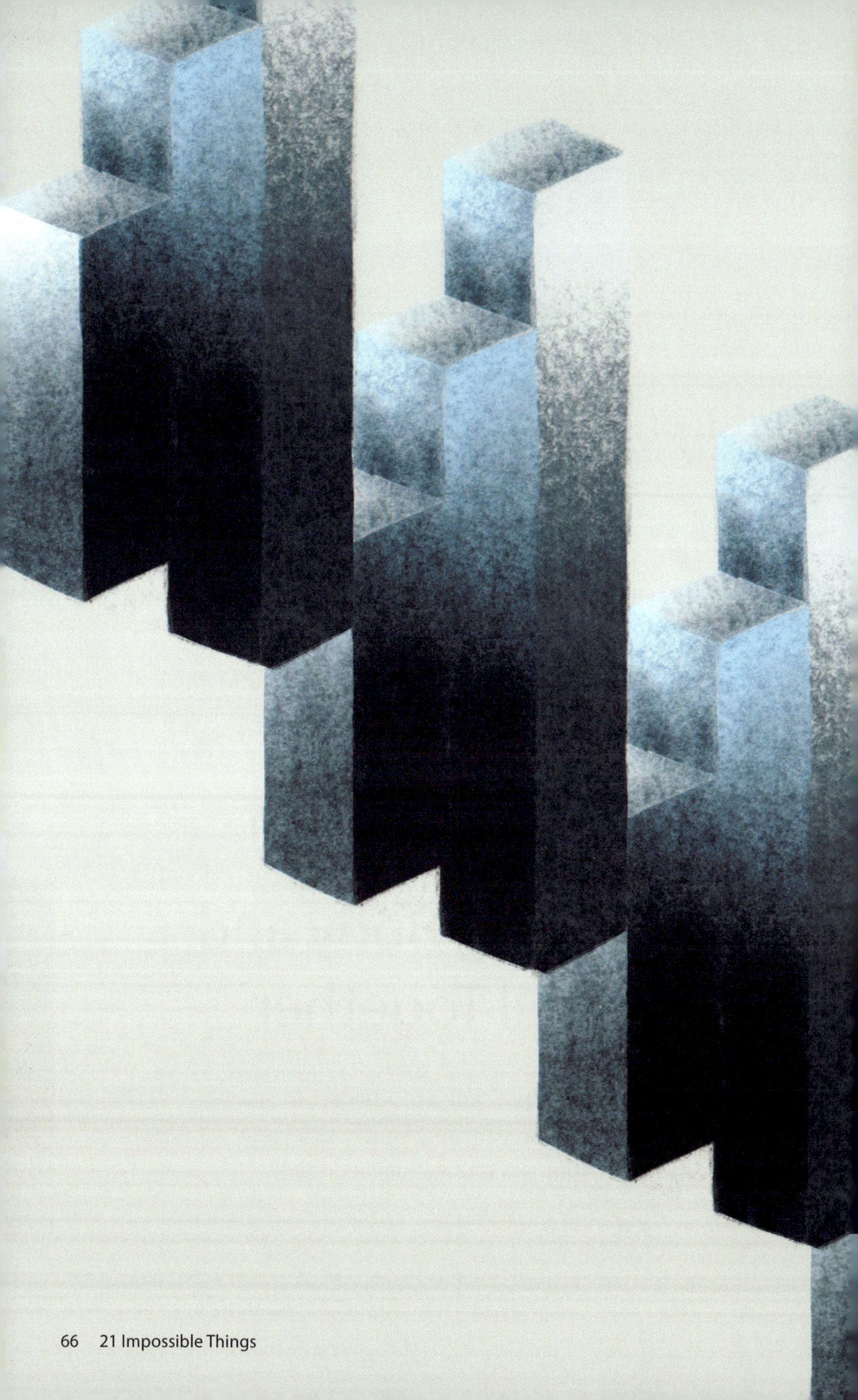

IN THE NETHERLANDS, 400 years ago, a soldier became a scholar—and never wanted to stop learning. French immigrant René Descartes (say it R'nay Dey-Cart) was nearly 40, but he continued to study at universities.

That was a good thing, since many of his ideas became foundational concepts for modern science. He worked on medicine, early robots called automatons, and other things. However, philosophy became his main interest.

At the time of this story, one thing distracted him from his work: love.

He was renting a room from a bookseller, and he had fallen in love with Helena, the bookseller's housemaid.

She loved him too. Eventually, they wed and had a baby, Francine, and the little family moved to France.

Then, tragedy struck. Shortly before her 6th birthday, Francine fell ill and died.

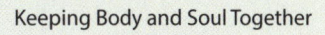

Keeping Body and Soul Together

René Descartes adored his child. He was broken-hearted and wept openly— unusual for a man in that society.

He switched his focus from medicine to deeper questions: What is a person? Where has Francine gone? What is reality, and how exactly do human personalities, like my child, fit into it?

This took him back to some ideas he had developed earlier. He'd read that Plato and Aristotle believed that the soul drove the body like a captain piloted a ship.

He decided that the pilot analogy was wrong. If your body is hurt, you feel physical pain. But if a ship is damaged, the captain isn't physically injured in the same way.

The opposing idea that things and people were *both* just physical objects also seemed wrong; it implied that there was no such thing as a soul. Francine's dead body felt like a child-shaped shell that had been emptied of something. What had gone?

These were humanity's big questions: How does consciousness fit into physical reality? Do both exist?

Remember what the Eastern mystics said about dreams in Story 5? That what you see and hear and smell and feel may not be "true" at all?

René Descartes decided that the real question that thinkers should be asking is not "What seems true?", but "What can I be certain of?"

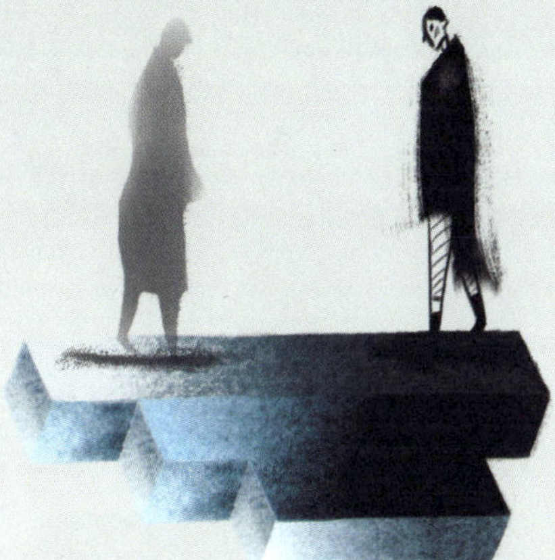

21 Impossible Things

And the answer was: only one thing.

That thing was thought. By that, he meant the consciousness that caused your thoughts to occur.

You think, and therefore you exist.

All other "proofs" that physical reality existed should be discarded, since reality could be a dream or an illusion or a projection, as Vasubandhu and Zhuangzi had proclaimed.

Descartes became most famous for his brilliant idea which was just five words long in English, "I think, therefore I am."

This wise Frenchman is often called a "duallist", suggesting that he divided reality into two things. But if we read his works, we get a different impression. He believed in physical things and mental things, and a mysterious third element where both met: that element was the human consciousness.

Some people say that René Descartes had a life-sized automaton of little Francine made, and travelled with it wherever he went—before sailors in a storm-tossed ship found it in his luggage and threw it overboard in a desperate bid to dispel bad luck and avoid drowning.

There's no proof that incident happened, although it does combine many of his interests.

In the following centuries, philosophers and scientists realized that many aspects of "the mind-body problem" remained one of humanity's greatest unsolved mysteries.

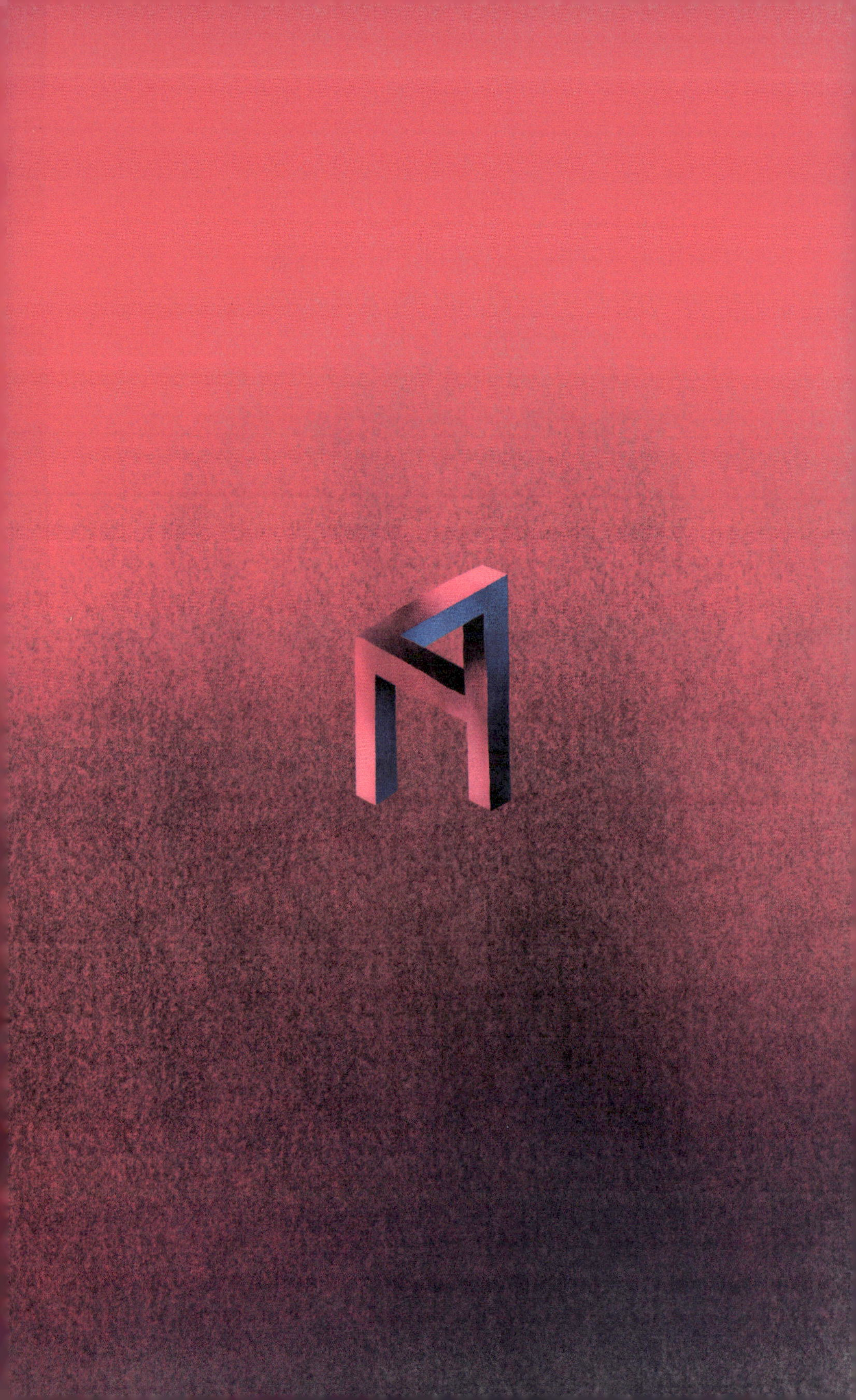

10

You Can Move an Elephant

Two mysterious and widely misunderstood forces, both crucial to understanding reality, are hidden in plain sight, Isaac Newton realized.

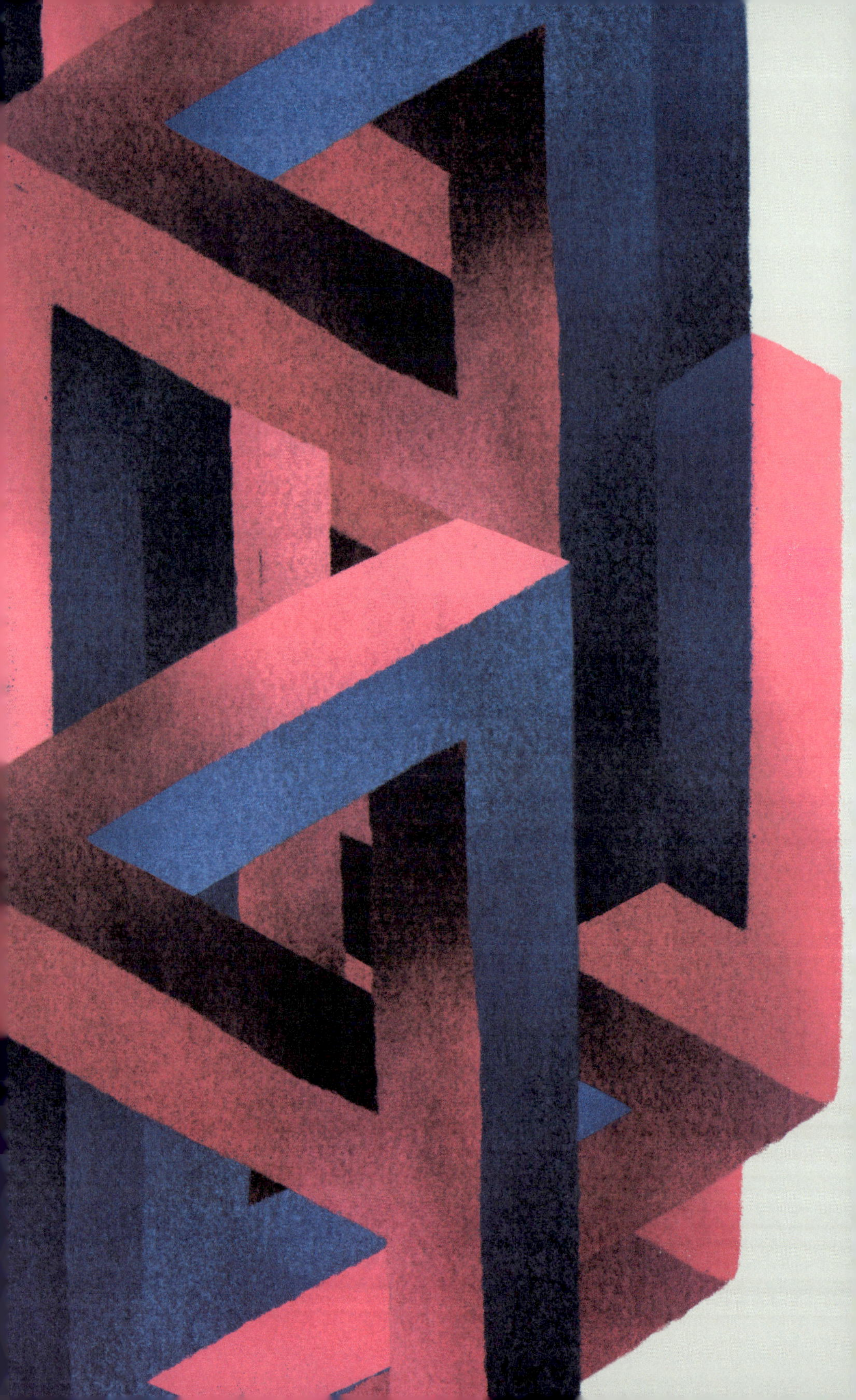

IMAGINE AN ELEPHANT on a wheeled platform. You and I have the job of pushing it to the jungle, a kilometer away. Hard work, for sure!

Or is it? Perhaps not.

To learn why, we can consult Isaac Newton, a man born in the year Galileo died, and who had read a book written by René Descartes.

Now, if you look up Newton on the Internet, you often see a cartoon of a man sitting under a tree: an apple falls on his head, causing him to "discover" gravity.

This is a well-known story—but it isn't true. The real story is less dramatic in one way, but far more dramatic in another.

Thinkers such as Aristotle and Galileo already knew that the world was round and that there was a force that seemed to pull everything towards the center of the Earth.

They already knew that apples fall from trees towards the center of the planet thanks to this force, which Aristotle called "convergence".

But what about the stars? Why did they not fall towards the center of the Earth? Aristotle believed that a mysterious force called quintessence (a Latin word meaning "fifth element") had been made by the Uncaused Cause to hold the stars in space. Galileo claimed that God caused things in the heavens to be held in place. Today, we say that a mysterious, indefinable element called "dark energy" holds the stars in space. These ideas are not very different.

You Can Move an Elephant

Newton's breakthrough was this. He was thinking about the moon, and puzzling as to why it didn't fall to Earth, when the apples on his trees in his garden came to mind. They would fall straight to the ground — unless you plucked one and threw it!

Newton realized that there were TWO separate forces that caused things to be how they were.

We could call the first force gravity, and it seemed to pull objects downwards or inwards.

And the other was inertia, which caused things to stay still if they were not moving, or keep moving if they weren't still. That last part of our understanding of inertia is crucial. (Remember Galileo's eternal football, flying through space forever in Story 7?)

The force of gravity pulled at the moon, trying to make it crash to Earth. But the force of inertia kept the moon flying through space.

Both these forces were perfectly balanced — causing the moon to circle the Earth continuously.

The same perfect balance of the exact same two forces caused the Earth to orbit the sun without crashing into it.

You Can Move an Elephant

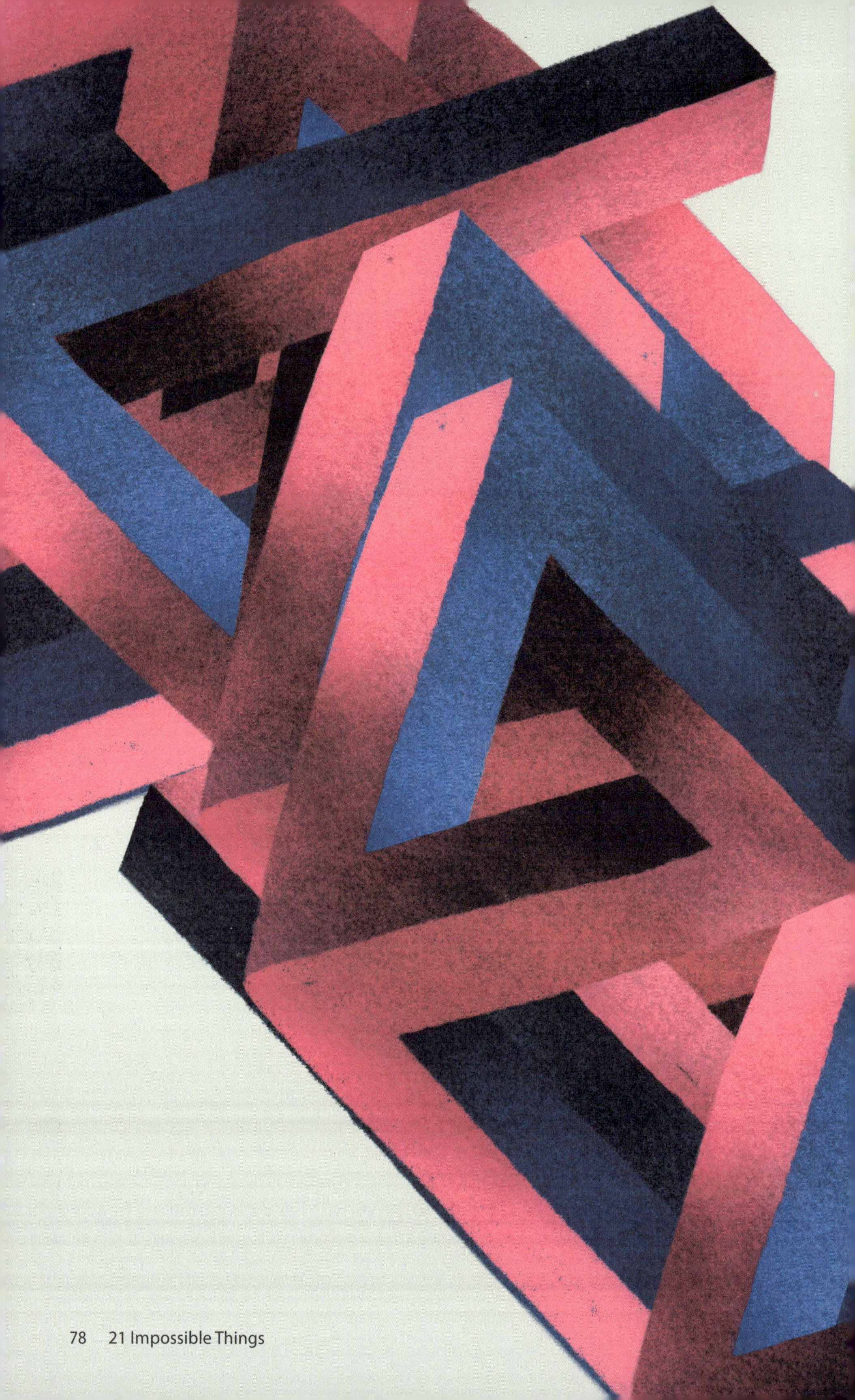

In fact, Newton realized, an enormous series of impossibly perfect balances caused the entire universe to exist as it does! Everything was incredibly delicately balanced. There must be a Mind behind it all, he thought. Everything was too carefully fine-tuned to be a product of random factors.

Now, back to our elephant-moving challenge at the beginning of this chapter.

It takes a lot of strength to get our elephant on a trolley moving. But once we get it rolling, the inertial force takes over and (as long as the road doesn't suddenly slope upwards), the trolley should move along by itself with no need for us to push at all!

Problem solved, thanks to the remarkable force that is hiding in plain sight: inertia—which causes things to stay still if they are not moving, or keep moving if they are not still.

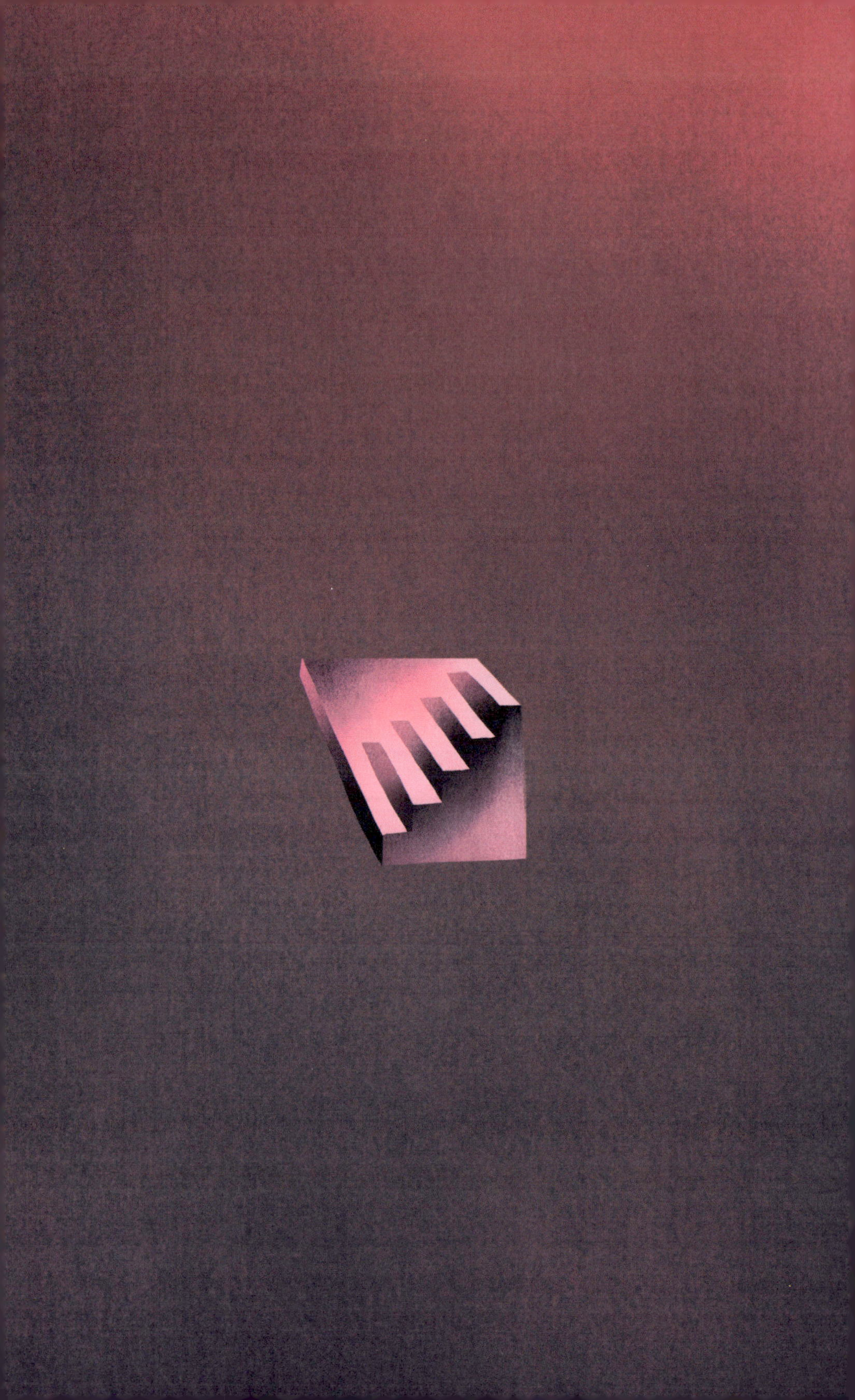

11

Invisible Hands, Everywhere

Isaac Newton is famous as the "discoverer" of gravity: but his letters show that he believed that what he described could not exist!

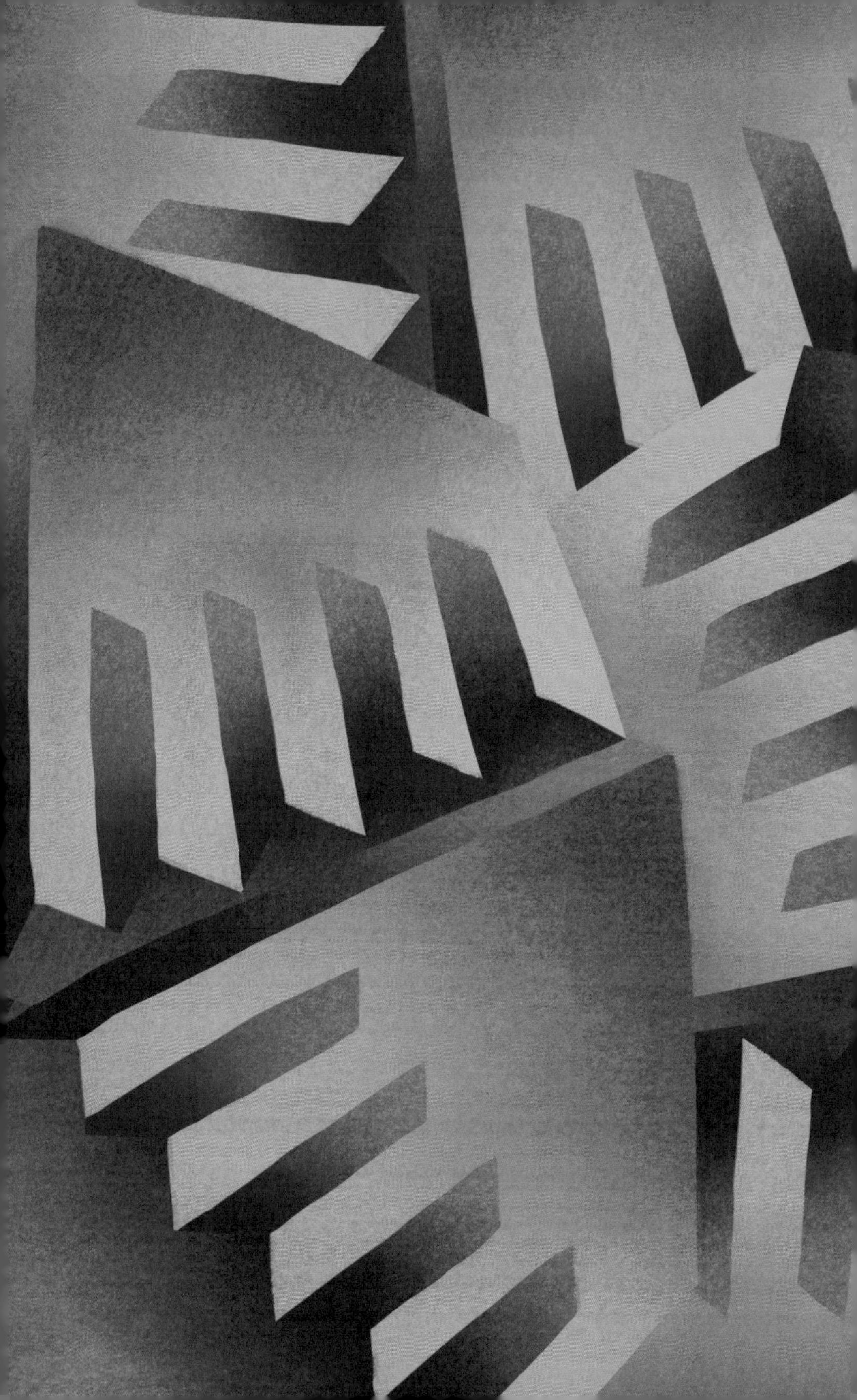

IT'S ACTUALLY QUITE funny that so many textbooks and websites identify Isaac Newton as the discoverer of gravity.

He didn't believe in it.

And what's more, he thought that intelligent people (any individuals with "a competent faculty of thinking") could never be tempted to accept such a silly notion.

Picture him sitting in his writing room in the year 1692, puzzling over the problem.

The issue was this: Having read the works of Aristotle and Galileo and others, Newton knew that a force seemed to pull things towards the center of objects, which is why we don't fall off the planet.

He himself had made a huge contribution to the subject by naming the force "gravity". He even wrote a simple formula that showed how it kept the moon in the sky, and quantified its effects.

Invisible Hands, Everywhere

Thanks to Sir Isaac Newton, today we can make detailed, accurate calculations about gravity that are crucial for the age of airplanes and spaceships.

But his triumph back then hid one huge, important problem: Gravity, as described, could not exist.

Gravity spread through empty spaces that didn't even contain air, and were perfect vacuums. That was not possible.

It worked on distant objects without touching them in any way – which was not supposed to happen.

Gravity worked like invisible hands pulling things towards the center of objects—but if those hands existed, you'd be able to detect them in some way. Therefore it seemed certain that they weren't there.

Gravity was "transmitted" from object to object, such as the sun and the moon and the Earth, without any medium. Sound waves need air, waves need water, but gravity travels through nothing. That made no sense to him at all.

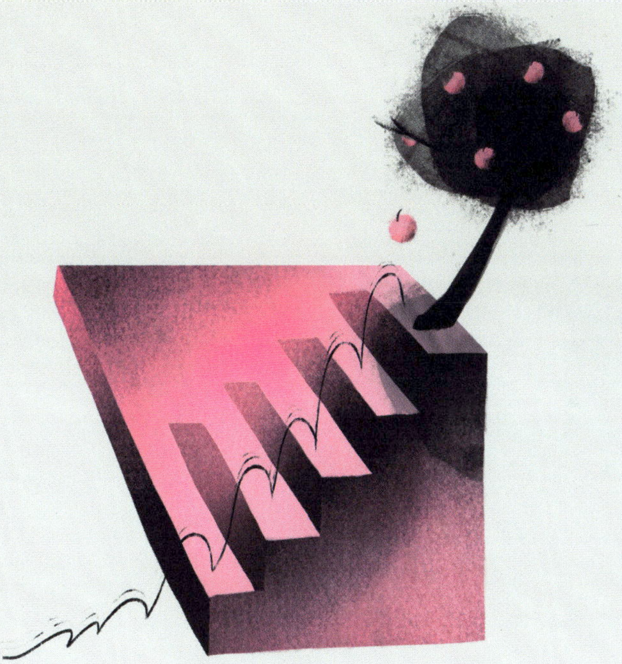

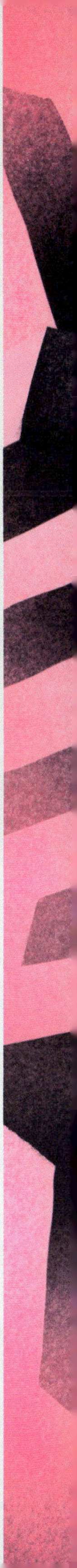

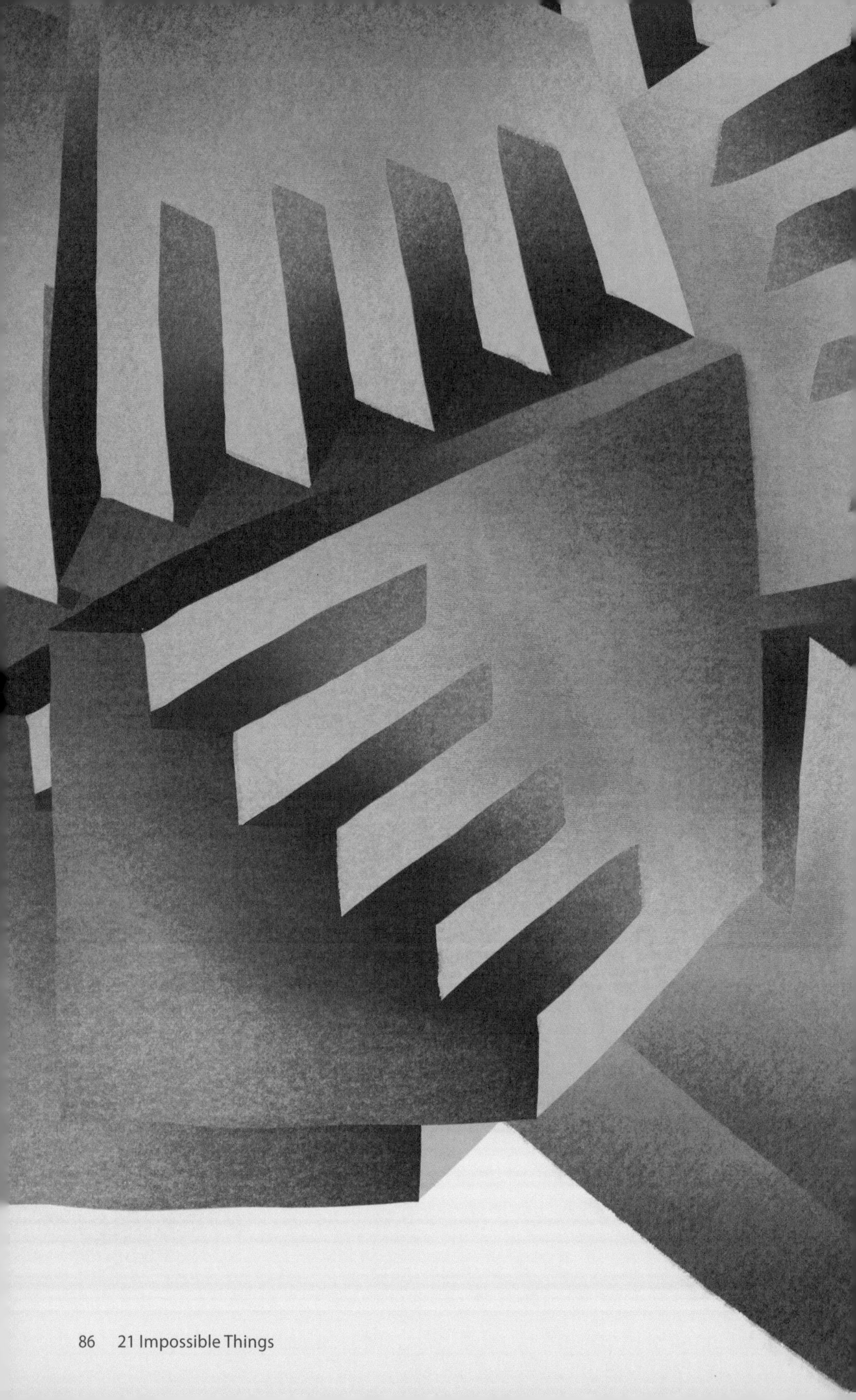

Oh, the effects of gravity were real enough, but no smart thinker could accept the existence of an invisible pulling force, he felt.

Newton sat down at his writing desk and shared his worries in a letter to his friend Bentley.

"That one body may act upon another at a distance through a vacuum without the mediation of anything else" was "an absurdity," he wrote.

There was a deep mystery here, and he became convinced that there were "causes hitherto unknown" that might one day be uncovered as fundamental to the "phenomena of nature".

Almost 200 years after his death, the mystery of gravity would be at least partly solved — and we would learn the answer.

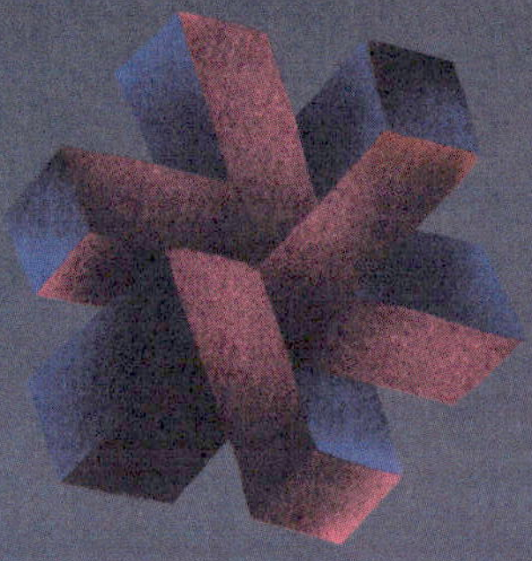

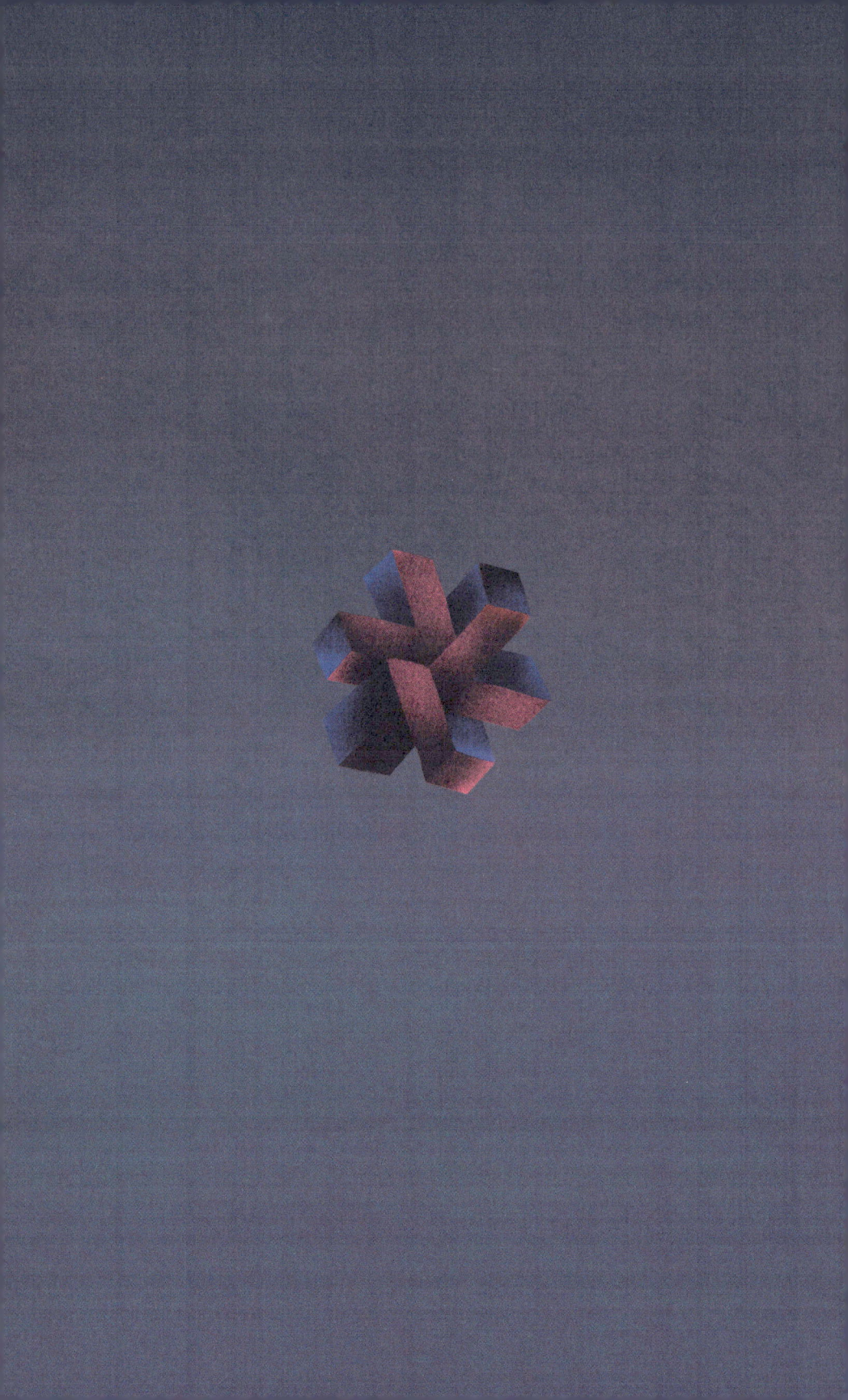

12

Is This the Real Life?

What if things only become real when you're looking at them? The thought of this sounds impossible to most people.

GEORGE BERKELEY (say it "BARK-lee") was a young university teacher in Dublin, Ireland, 300 years ago.

He was a deeply religious man (who ended up a Bishop) and one of the early champions of a way of thinking called empiricism — this means that you take your beliefs from observations, not from purely thinking about things. Empiricism became a foundational concept of modern science.

Berkeley read about an interesting experiment by a man named John Locke.

The experiment goes like this: you put your left hand in a bucket of hot water and your right hand into a bucket of cold water. Now you put both hands at once into a bucket of lukewarm water. The left hand will send you a message that the water in the third bucket was cold, and the right hand will send you a message that it was hot!

Which hand is lying? Neither. So they are both sending accurate messages, despite giving you opposite conclusions.

This shows that an individual's understanding of reality comes from perception rather than empirical facts. Both are important, but they must be recognized as two processes, each of which could be applied to the other.

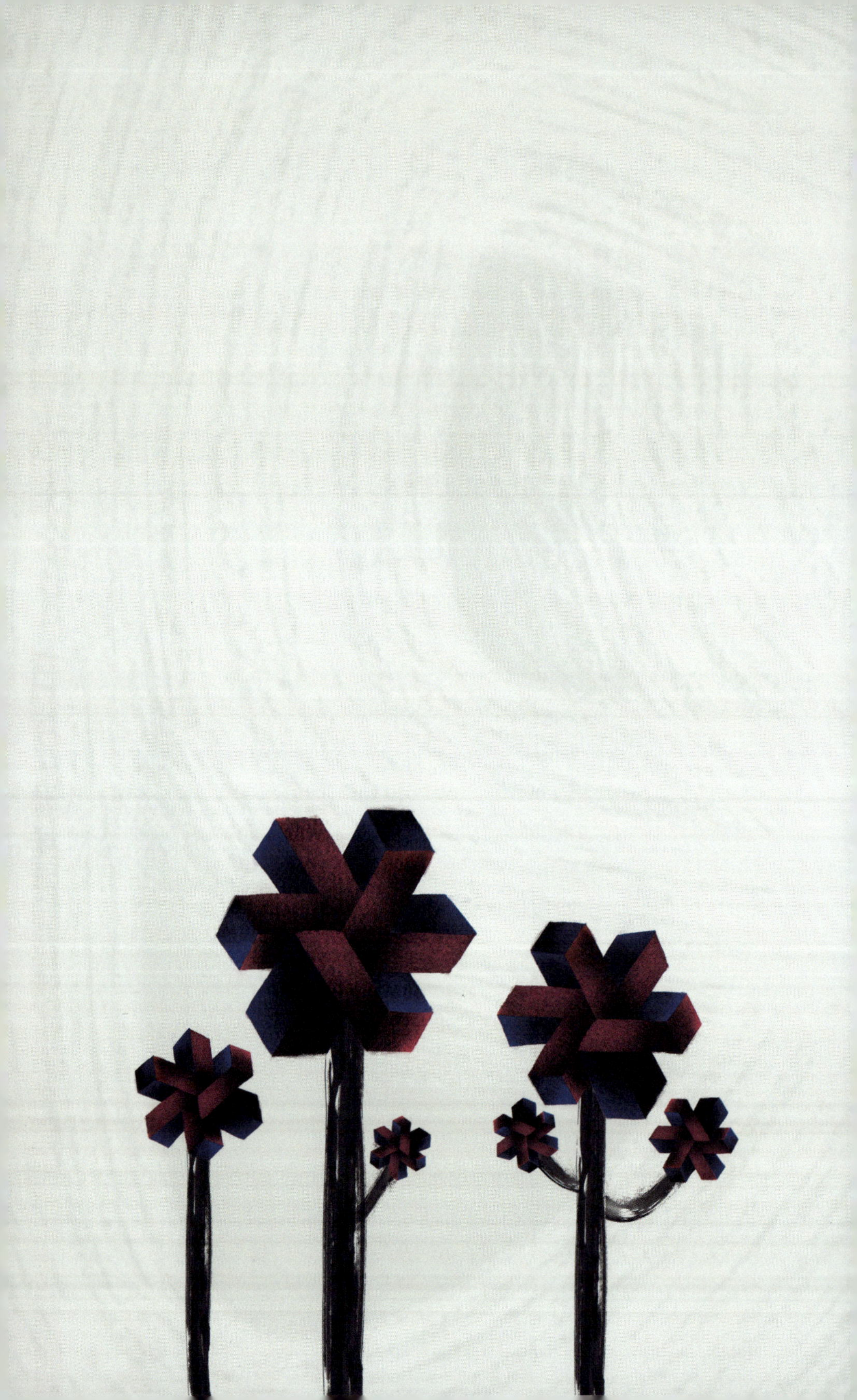

George Berkeley came to believe that empiricism showed that all of our understanding of reality came from perception. In one of his books, he posed a question: If a forest of trees existed without animals or humans looking at it, could it have never existed at all? After all, there were no conscious beings who were aware of it. If matter arose from mind, then a place without minds, might be a place where there was no matter.

Perhaps reality was observer-dependent, he thought. That means that things only exist when they are being looked at.

This sounded like a crazy idea, and in all honesty, many people were critical of him. One famous writer at the time said: "I refute it thus" and kicked a stone to show that things were solid.

But today we can think of Berkeley's view of reality like the world inside computer games – depending on which way you scroll, the processor creates an environment for the character to step into.

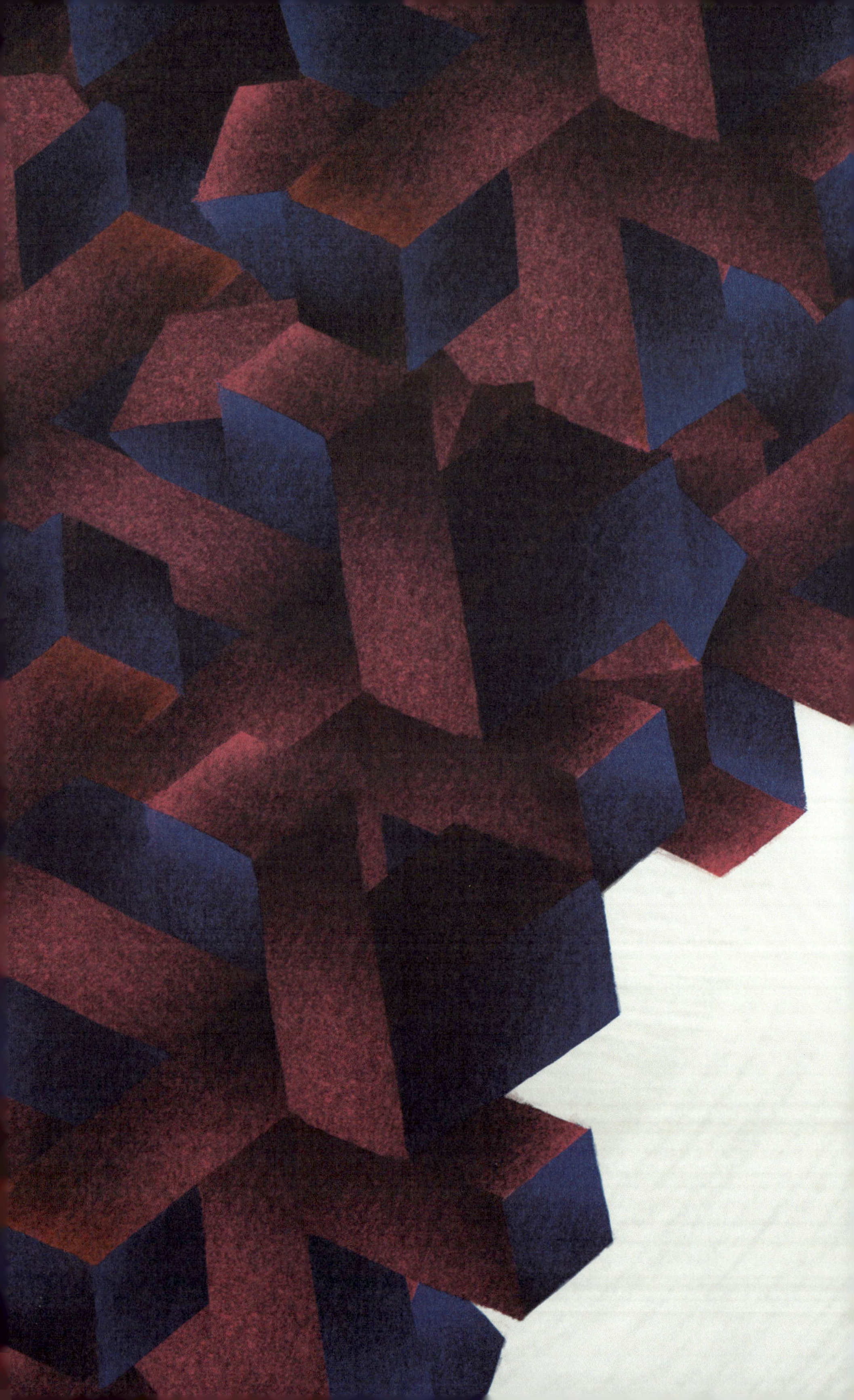

People wrote down George Berkeley's puzzle as a short, neat question: "If a tree falls in a forest and there is no one to hear it, does it still make a sound?"

But is reality really so? Surely it makes no difference whether someone is looking at the forest or not.

So it would seem. Yet Berkeley was right about so many other things that this idea of his, wacky as it sounded, was remembered and discussed endlessly.

More than 200 years later, a series of empirical experiments would reveal extraordinary results that threw doubt on the idea that reality existed in a way that was independent of observers. People like Albert Einstein and Niels Bohr would remember the words of the Irish priest. "Is the moon not there when we're not looking at it?" Einstein asked Bohr, after reviewing the Dane's work. It was a serious question.

It seems crazy, but scientists had good reason to think that the obvious answer — "Of course it is" — might be wrong.

Today, many if not most theoretical physicists think Berkeley was right. Physical reality is an illusion. Matter emerges, somehow, from mind. That's baffling, right? But hold on — all will become clearer later.

Is This the Real Life?

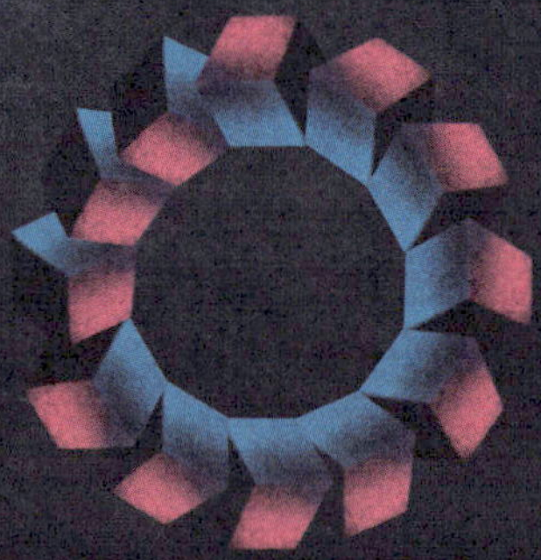

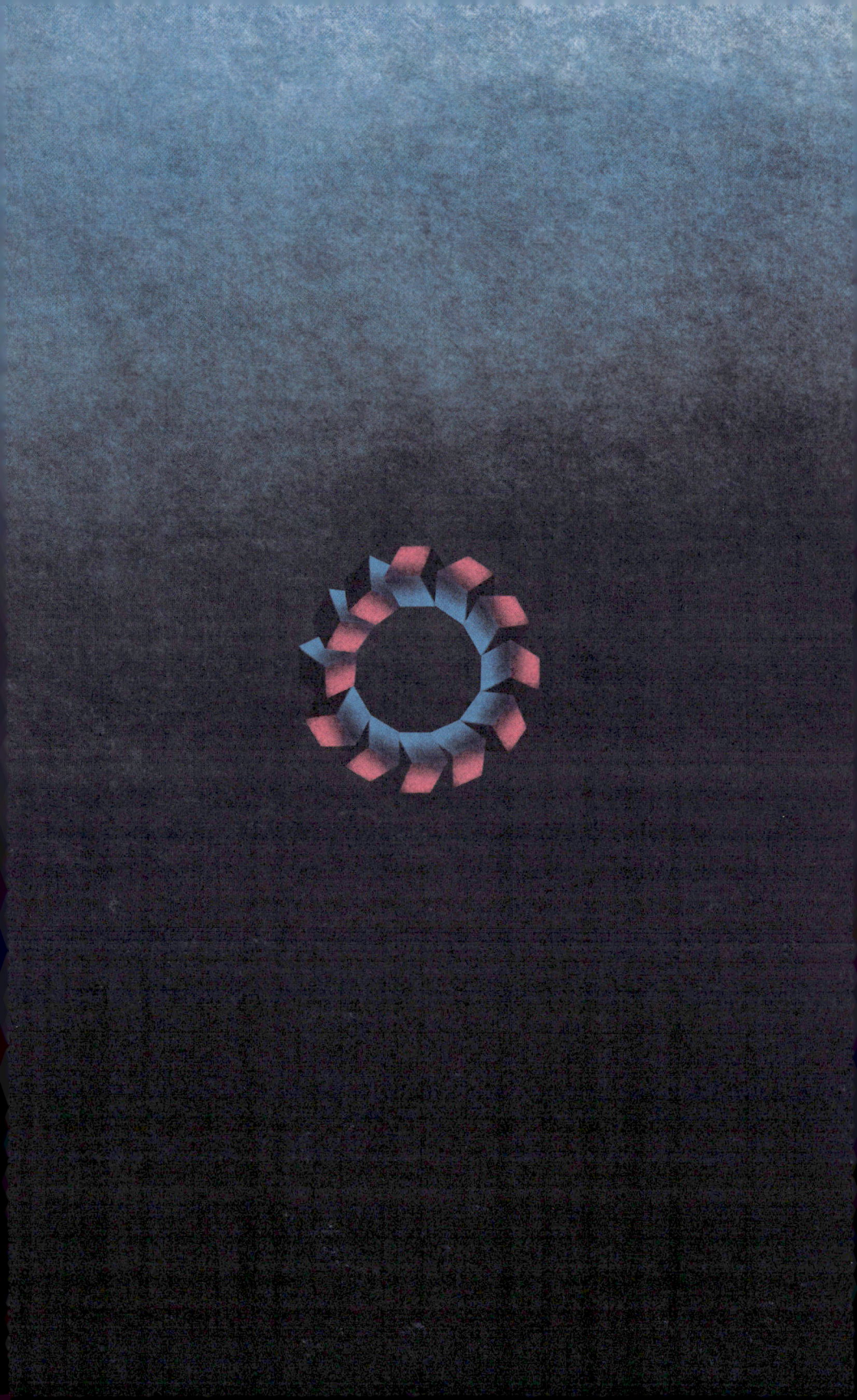

13

The Time Traveller

Time travel is a fun idea for stories, but it would be impossible for us to work out how it could happen in real life — wouldn't it?

HUMANS MAKE MISTAKES — and then wish we could go back in time and undo them. But of course we can't.

Time travel is popular in fiction, but not in the real world, it seems.

In the West, the best-selling stories of the 1800s were hyper-imaginative. Frenchman Jules Verne wrote about fantastic voyages. Englishman Charles Dickens wrote about a miser named Scrooge who was taken back in time to see his past. American author Washington Irving wrote about a man who slept for 20 years and woke up feeling like he'd travelled to the future. In the 1880s alone, at least eight tales were written about time travel, including an 1888 story by a British author named Bertie Wells.

But another idea was circulating, too. By that time, many people understood that we live in a three-dimensional world — but became fascinated by stories about what could exist in a parallel universe that some called "the fourth dimension".

Maybe that's where ghosts come from?

Maybe it's another world we can enter from this world?

Maybe that's where aliens come from?

A more scientific author, a mathematician named Charles Hinton, thought about this much more seriously and came out with a different idea.

He proposed that the fourth dimension was time.

To fully describe an existing object, he said, you'd have to say what it looked like in three dimensions, plus how long it existed for.

In 1895, Bertie Wells lifted Hinton's idea and used it to add scientific weight to another time travel story.

In The Time Machine, a character called The Time Traveller appears and asks his audience a rhetorical question: "Can a cube that does not last for any time at all, have a real existence?... Any real body must have extension in four directions: it must have Length, Breadth, Thickness, and — Duration."

The Time Traveller

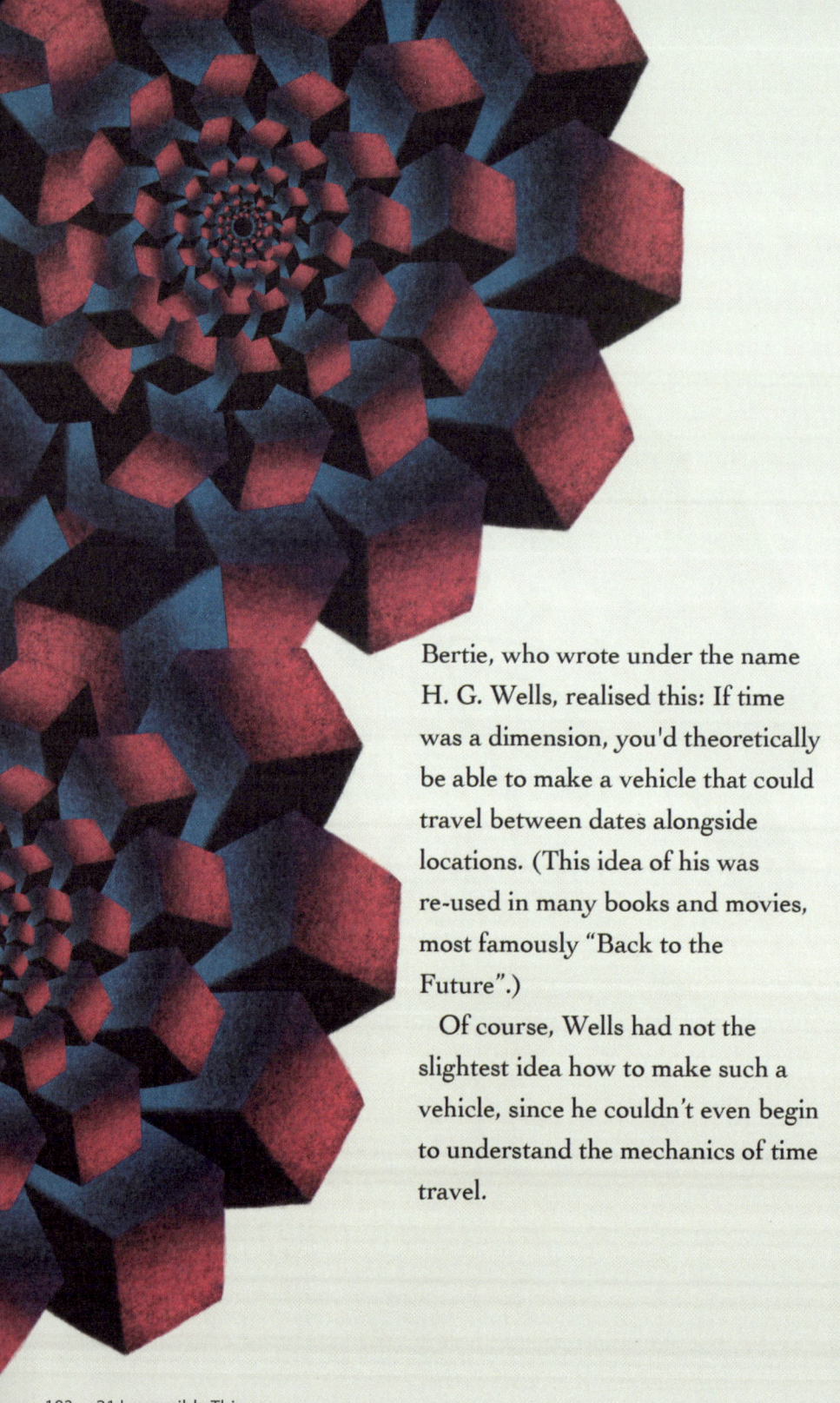

Bertie, who wrote under the name H. G. Wells, realised this: If time was a dimension, you'd theoretically be able to make a vehicle that could travel between dates alongside locations. (This idea of his was re-used in many books and movies, most famously "Back to the Future".)

Of course, Wells had not the slightest idea how to make such a vehicle, since he couldn't even begin to understand the mechanics of time travel.

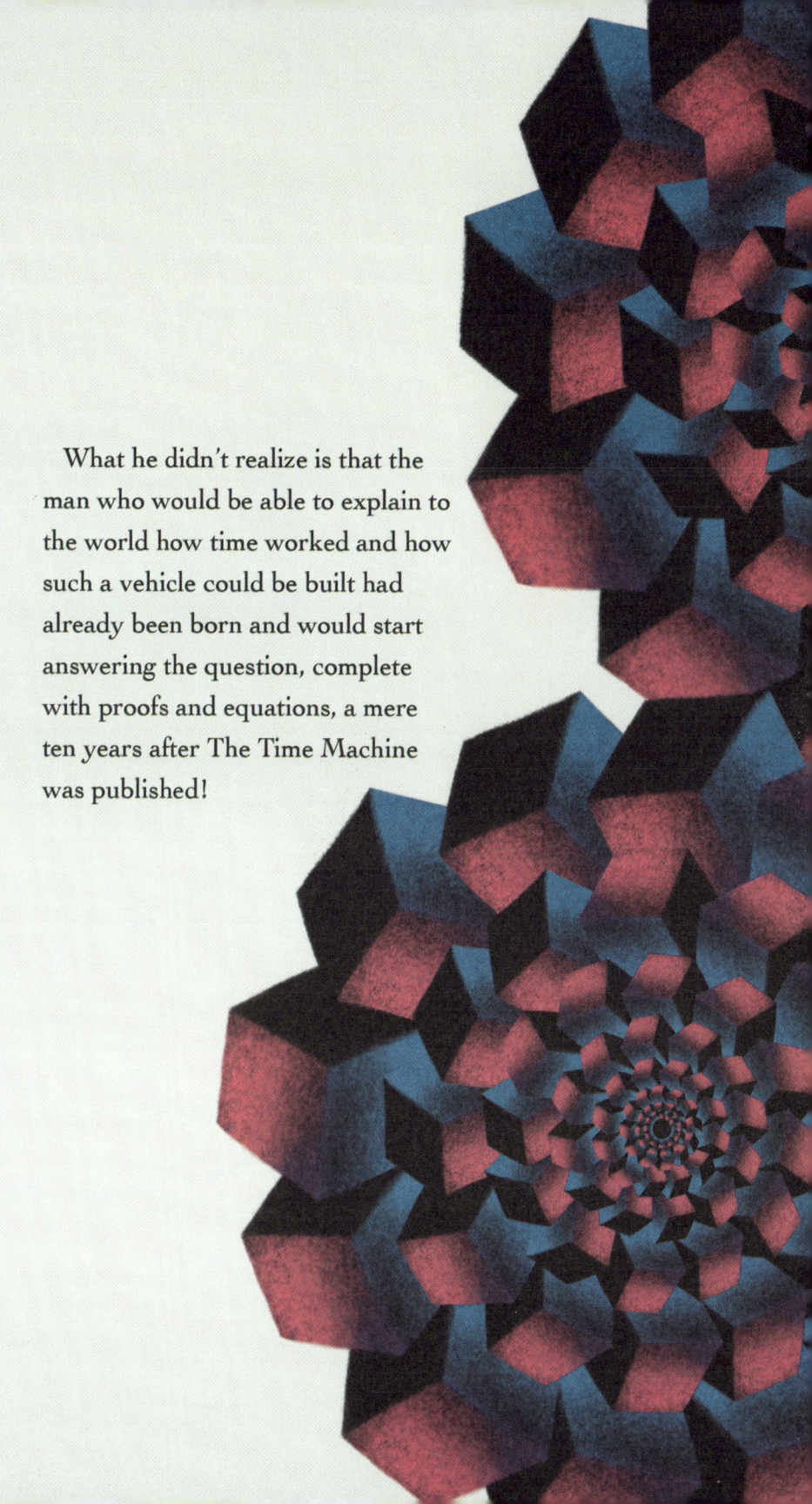

What he didn't realize is that the man who would be able to explain to the world how time worked and how such a vehicle could be built had already been born and would start answering the question, complete with proofs and equations, a mere ten years after The Time Machine was published!

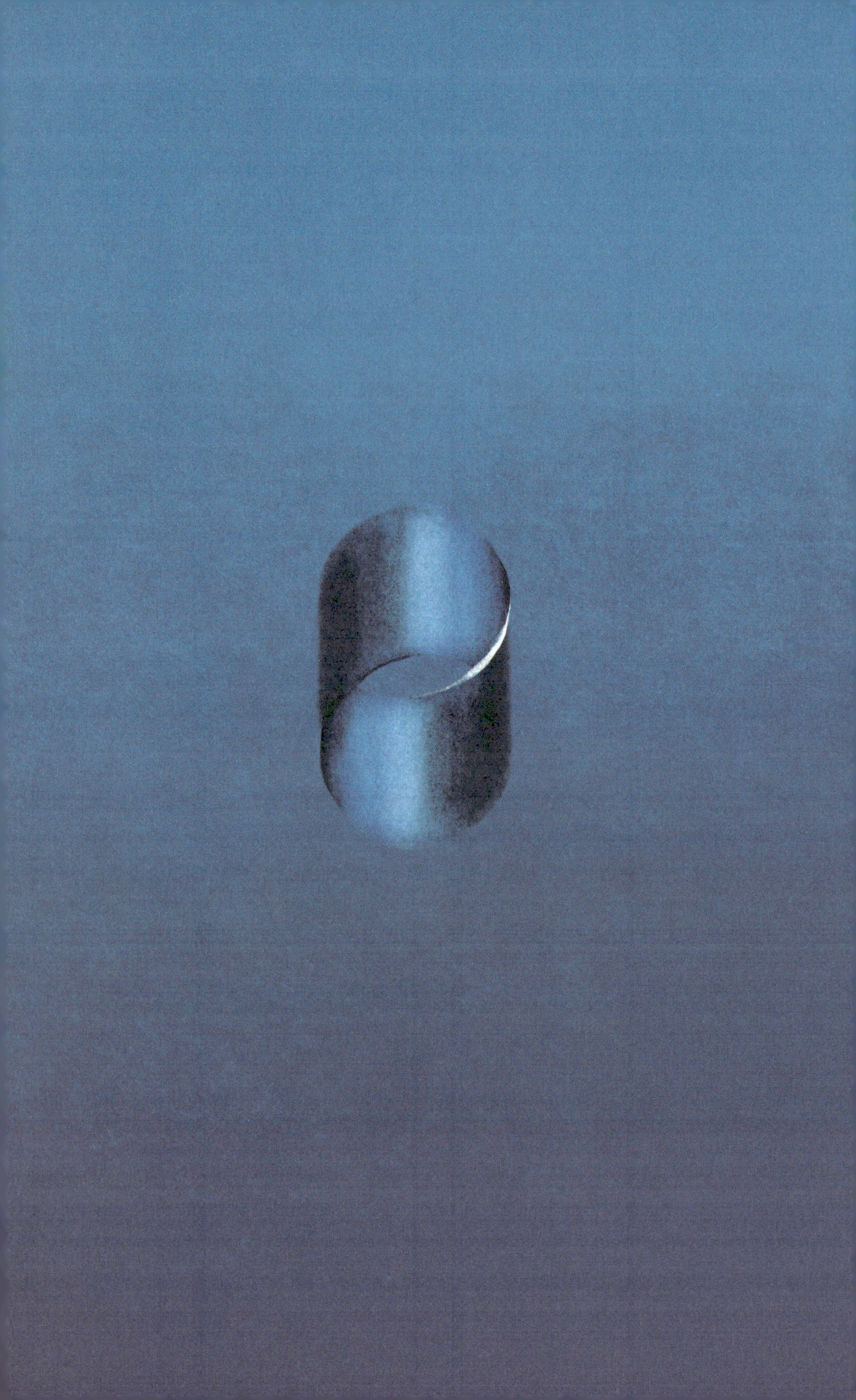

14

Time for a Change

Our rational minds tell us time is a process, but what if it wasn't? What if we instead thought of it as a place?

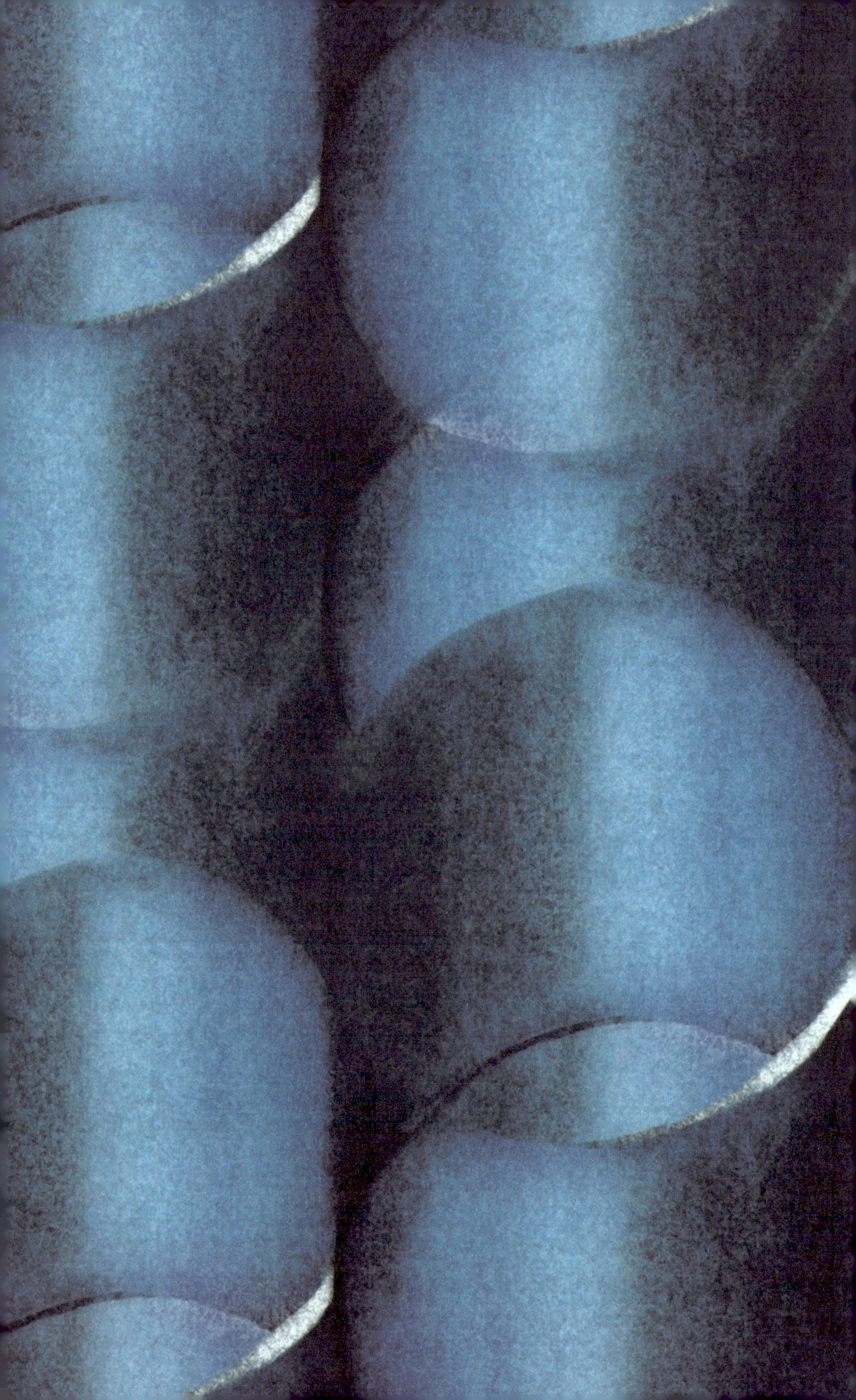

ALBERT EINSTEIN, A GOVERNMENT office worker in Switzerland, had a shock. He was thinking about the nature of reality when he realized something astonishing.

It was obvious that reality consisted of two things: physical reality (the space in which we live) and the passing of time (the unstoppable flow of seconds, minutes, hours and so on).

These two things together form the stage on which everything happens. We read about this in Story 4.

Space and time have always been the starting point at which all scientific and philosophical studies begin. If the universe was a ship in a harbor, space-time would be the anchor. If reality was a solar system, space-time would be the sun around which everything revolved.

This was obvious to everyone, so no one disputed it.

Except... *it was wrong.*

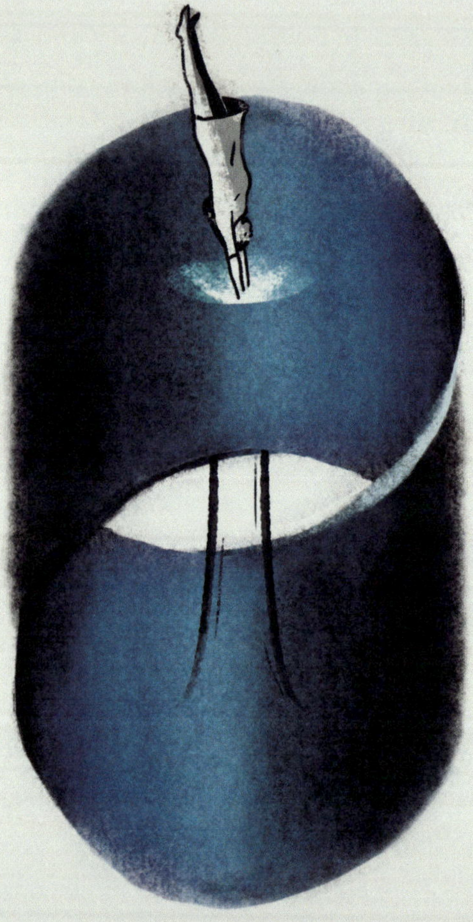

Einstein realized that the starting point of reality, the anchor, was something else: Light — an element that had been discovered to move at the speed of 300,000 km a second.

The truth is that light is the pivot of the physical world, and the number associated with it is not just the speed of light, but the speed limit of reality itself, Einstein realized.

Light is the foundation stone. Everything else adjusts itself to fit, including time and space.

What does that mean?

If the maximum speed of light is the unchangeable thing, the constant thing, the unalterable thing, everything else must be malleable — that means it can be altered, and in fact is altered, all the time.

The practical result is this: As light moves at its constant rate, time itself is sometimes bent or changed. And physical space is crushed or stretched or twisted.

Could we make use of this discovery in practical terms?

Yes. If we zoomed along in spaceships at high speeds, the way we experience time and space would change. If any of us actually reached the speed of light, 300,000 km a second, time would stop entirely! (But so would we — we wouldn't be able to move.)

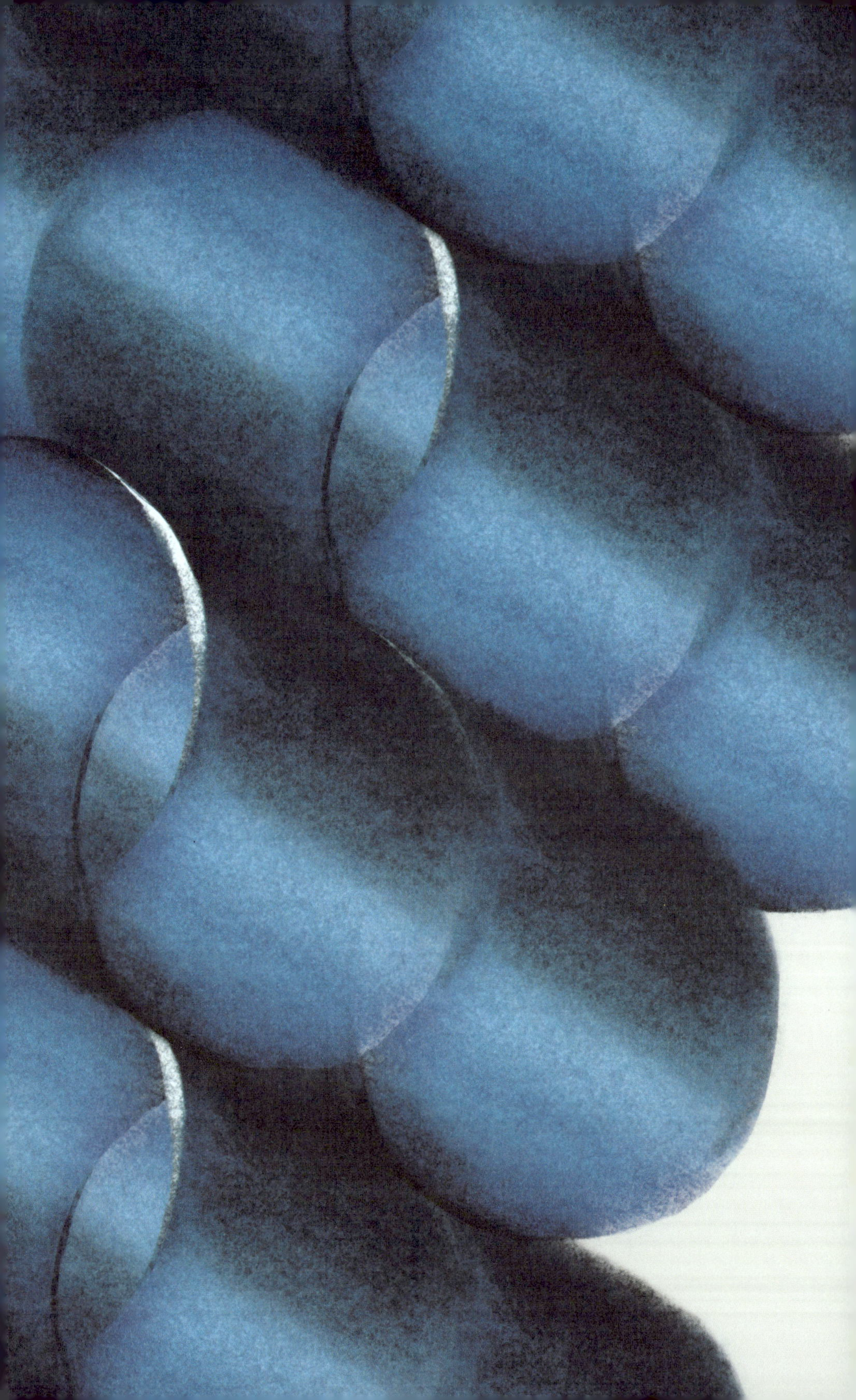

When Einstein wrote about this idea, it changed the way that humans saw reality itself — and sparked numerous other discoveries. Including this one:

How to Make a Real, Working Time Machine

1) Make a rocket that blasts you into space at half the speed of light or faster.

2) Note that because of your speed, the way you experience time and space will change. Time will move more slowly for you than the way the people you left on Earth will experience it, but it will feel completely normal to you, so you'd feel just fine.

3) Return to Earth just five years later — and you'll find yourself far in the future: the year might be 2050 or 2060 or later, depending on how far and how fast you travelled.

Because Einstein's idea focused on the way that things (like light and space and time) related to each other, people called it "The Theory of Relativity".

But it was really a cluster of ideas published in two sections, one in 1905 and the other in 1915-1916.

The idea we looked at in this story was just one of them — we'll look at the other in the next story.

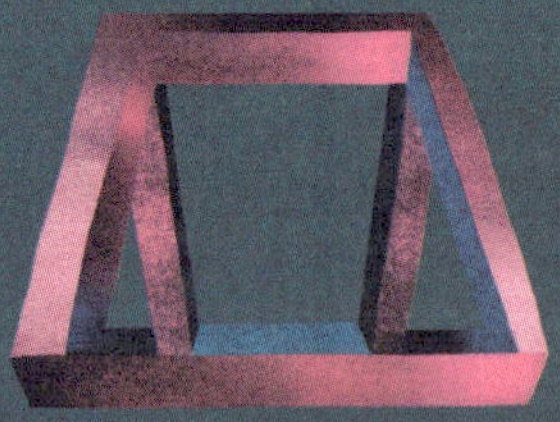

15

A Ripple in Reality

What if the space in which we live changes in all sorts of ways without us ever noticing? Actually, it's doing just that.

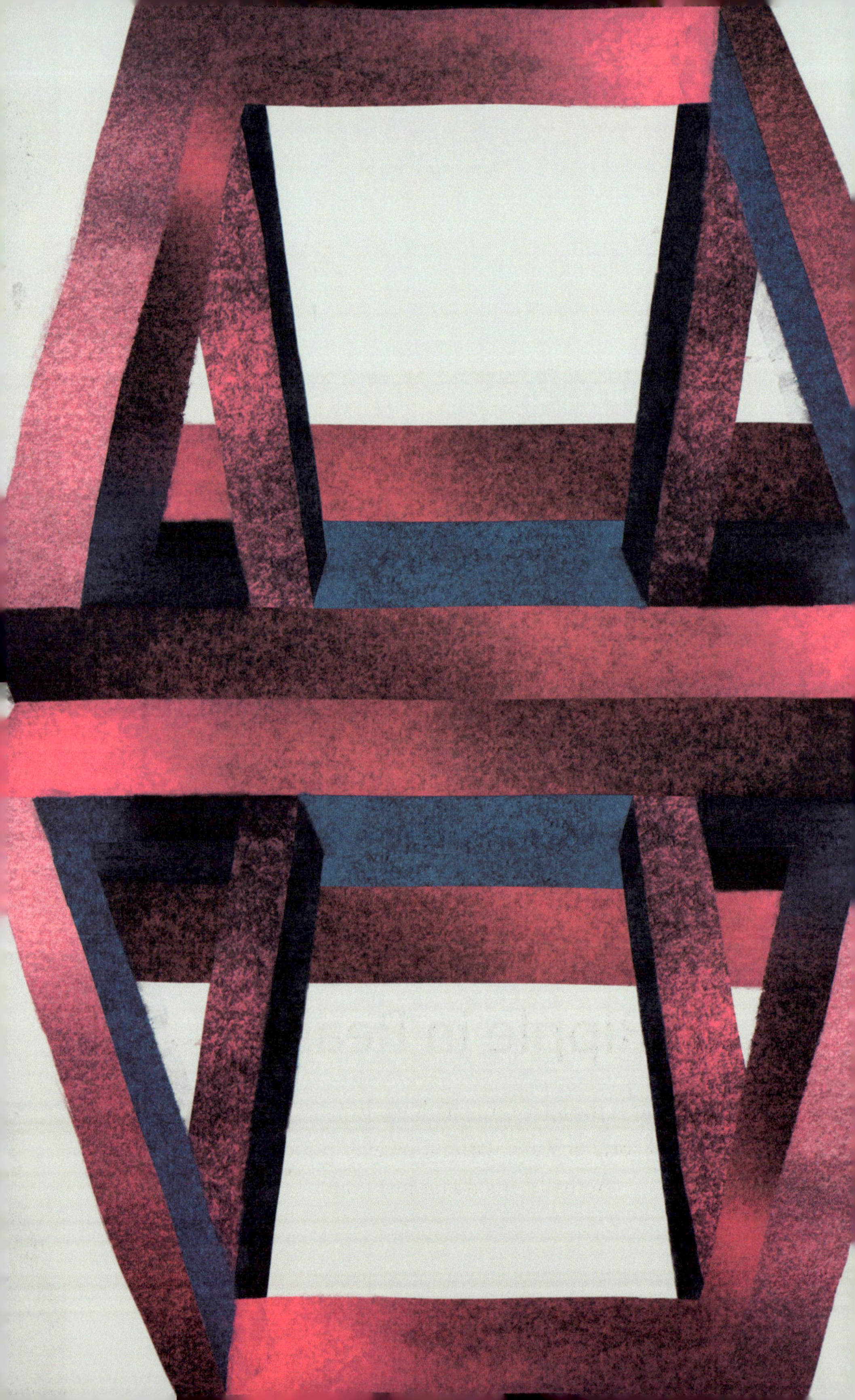

IMAGINE YOU HAD a giant kitchen sponge, bigger than a room.

Picture yourself in your backyard painting a scene on the side of that sponge. You draw a life-sized person standing on the floor level, and a light fitting at the top of the picture.

Just as you finish painting this, a real-life giant approaches your street.

The huge human being sits down on the roof of your neighbor's flat-roofed bungalow.

Through the window, you see the neighbor's ceiling bend down. She drops to her knees as the light fitting hits her head.

Luckily, she isn't hurt badly and crawls to safety.

Then the giant gets up and walks over to sit on your giant sponge, which looks like a more comfortable place to relax.

Will the figure you just painted be destroyed too?

The answer is no.

When the giant sits on the sponge, the image is squashed downwards but the elements stay in proportion. The gap between the character and the ceiling remains. The image springs back to normal when the giant leaves.

Now, is the reality in which we live like the neighbor in the bungalow? Or does it work more like the scene in the picture in the sponge?

Logic tells us that the first one is true.

But Albert Einstein realized that the second was true. Reality itself can be squeezed or stretched or rippled or bent or twisted into a curve or other shapes.

This is called the "curved space" theory.

If the reality in which we lived was squashed a bit, we wouldn't notice — because everything stays in proportion, including us. So ripples can run through reality itself. We call them "gravitational waves".

With this discovery, Einstein partly solved the mystery that Isaac Newton identified about how gravity works. (Remember Story 11?) As Newton realized, gravity seemed to pull at us with invisible hands, but couldn't actually work in that way.

What's really happening is that a big object like a planet distorts the reality around it, creating what we call a gravitational well. This holds us in place. No invisible hands needed.

A Ripple in Reality 117

Yet another part of Einstein's work focuses on the relationship between energy and matter. Remember in Story 4, ancient philosophers worked out that all matter (physical stuff) was made out of energy?

Einstein worked out the details. You've probably seen his famous equation, $E=MC^2$. That simply describes how energy and matter are related.

The "E" stands for "Energy".

The "M" stands for "Matter".

"C" stands for the never-changing speed of light (think of the word "Constant").

So matter multiplied by the speed of light squared (that is, multiplied by itself) is energy.

We think of the universe as a big empty space with stuff like stars and planets scattered in it. But really, it's just a big manifestation of energy — or, as the ancient Chinese philosophers said, Qi.

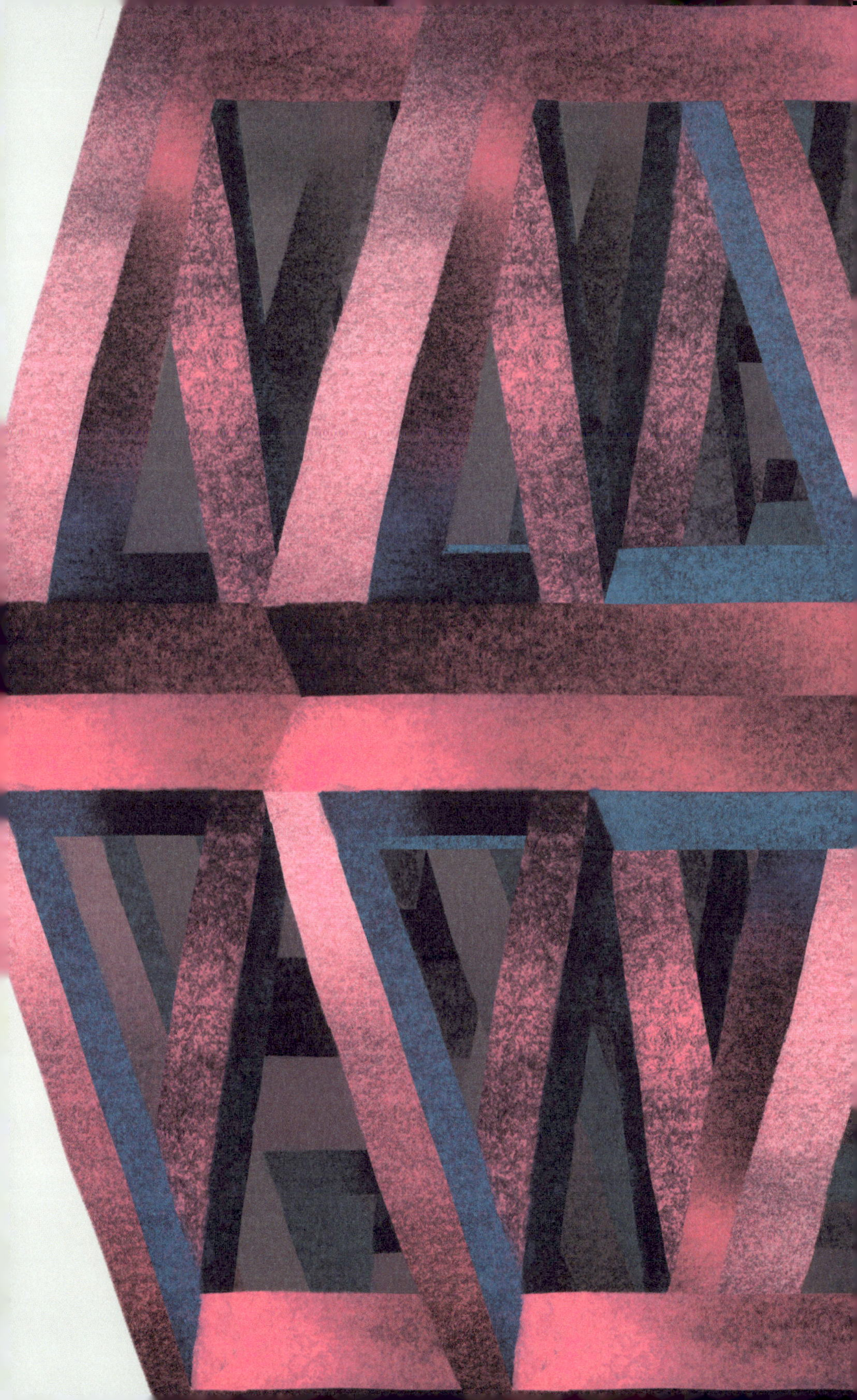

16

A World Made of Nothing

He tried to work out what was inside an atom, but the findings made no sense at all.

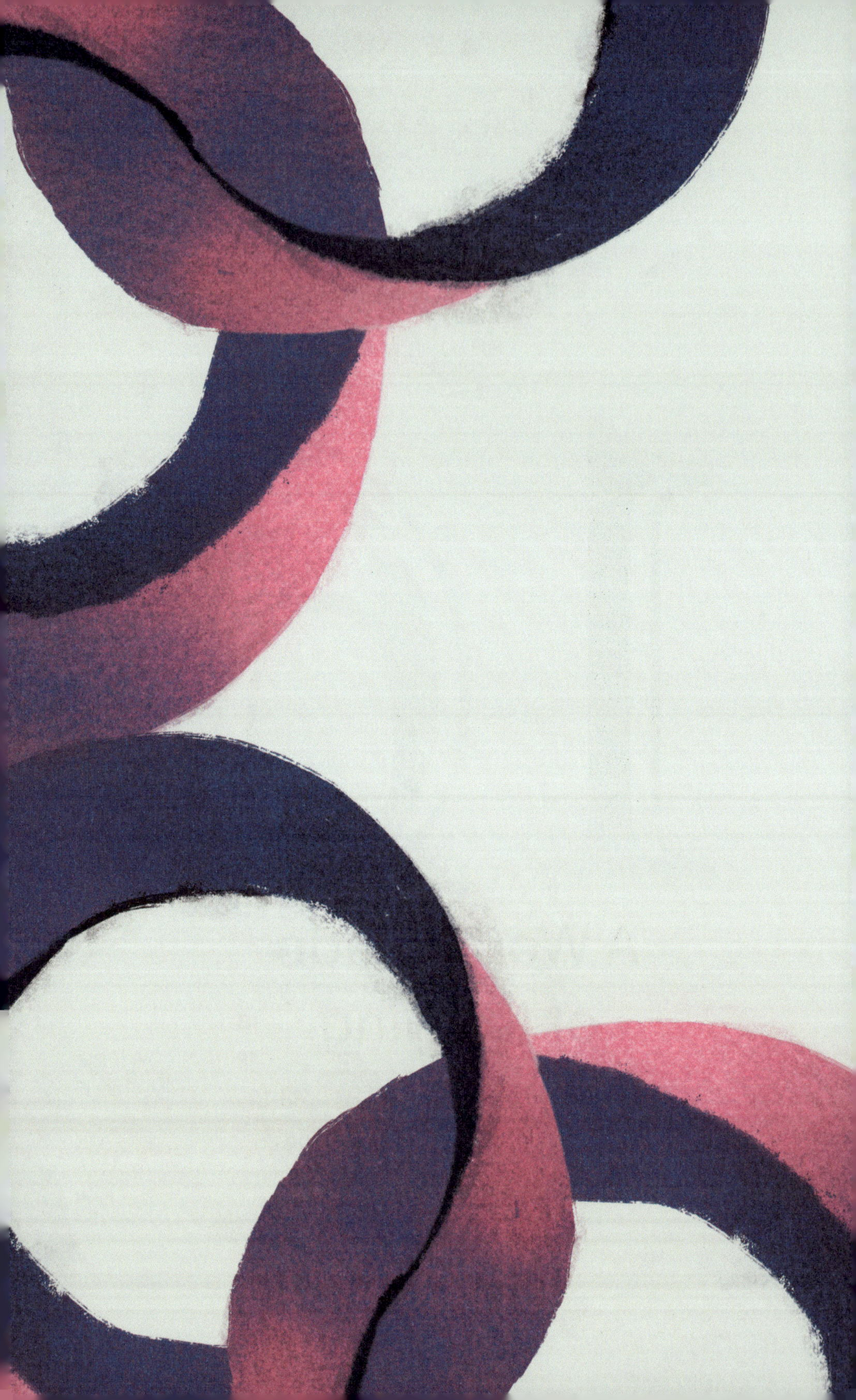

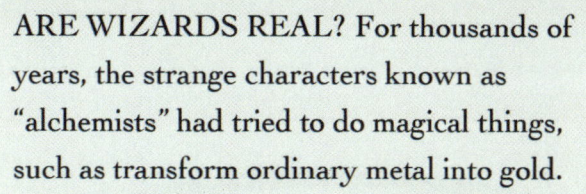

ARE WIZARDS REAL? For thousands of years, the strange characters known as "alchemists" had tried to do magical things, such as transform ordinary metal into gold.

But if you think about it, alchemists are just open-minded scientists. We call Isaac Newton a scientist, but he thought of himself as an alchemist.

Now, this part of our story begins in New Zealand, in the early 1900s. A little farm boy, who had later grown up to become a scientist, sailed across the world to spend time in Canada. He later moved to Cambridge University in the UK.

His name was Ernest Rutherford and he had no doubt that his "real" science work differed from the sort of things that alchemists or wizards did in storybooks.

And then one day, something extraordinary happened. He and his assistant watched as one substance transformed itself into another during an experiment in his lab. "Watch out, Soddy," he said to his assistant. "They'll have our heads off as alchemists!"

But the other members of the science establishment did not tease him. They were delighted by the achievement. It proved the theory that materials were made of different combinations and arrangements of atoms.

Democritus (Story 5) had been right about that.

Rutherford was given the Nobel Prize — but he didn't rest on his laurels.

The former farm boy decided that if solid things were made up of tiny balls, he should work out what the balls were made of.

Rutherford arranged for his junior colleagues to shoot extremely small particles at a thin sheet of gold. They were so tiny that they could penetrate the invisibly miniscule atoms of gold.

If the particles went through fairly smoothly, it would show that atoms were smooth inside. If some of the particles changed angle, it would suggest they were lumpy inside. The aim was really to see if Democritus's "uncuttables" really were solid, uncuttable balls, or whether there was something inside.

The results astonished Rutherford.

Most of the particles went right through, but a tiny number of them bounced straight back at the particle gun.

Rutherford's jaw dropped. "It was as if you fired a 15-inch shell at a sheet of tissue paper and it came back to hit you," he said.

Rutherford realized that the experiment's results as a whole showed what Democritus's atoms were made of nothing! They were balls of empty space surrounded by a mysterious soft boundary, but each one had something very tiny and very, very hard in the middle.

Rutherford named the strange thing in the middle "the nucleus".

He and the "nuclear scientists" who came after him eventually worked out that atoms were 99.9999 per cent empty.

And yet they made up everything in the world — in fact, everything in the universe.

That meant the universe was made of nothing!

The discovery caused mouths to drop open. The deeper we dug, the more mysterious the world was becoming.

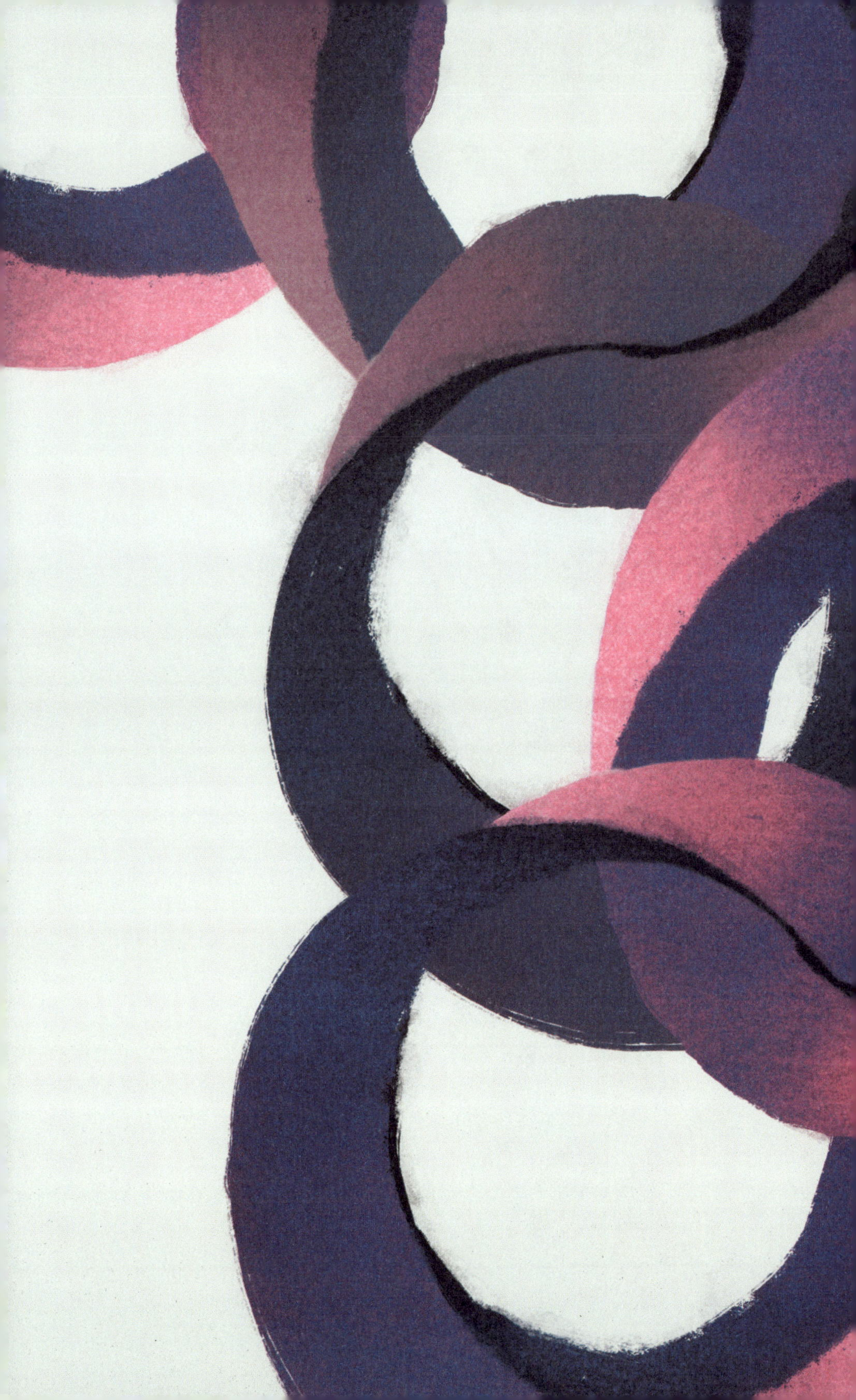

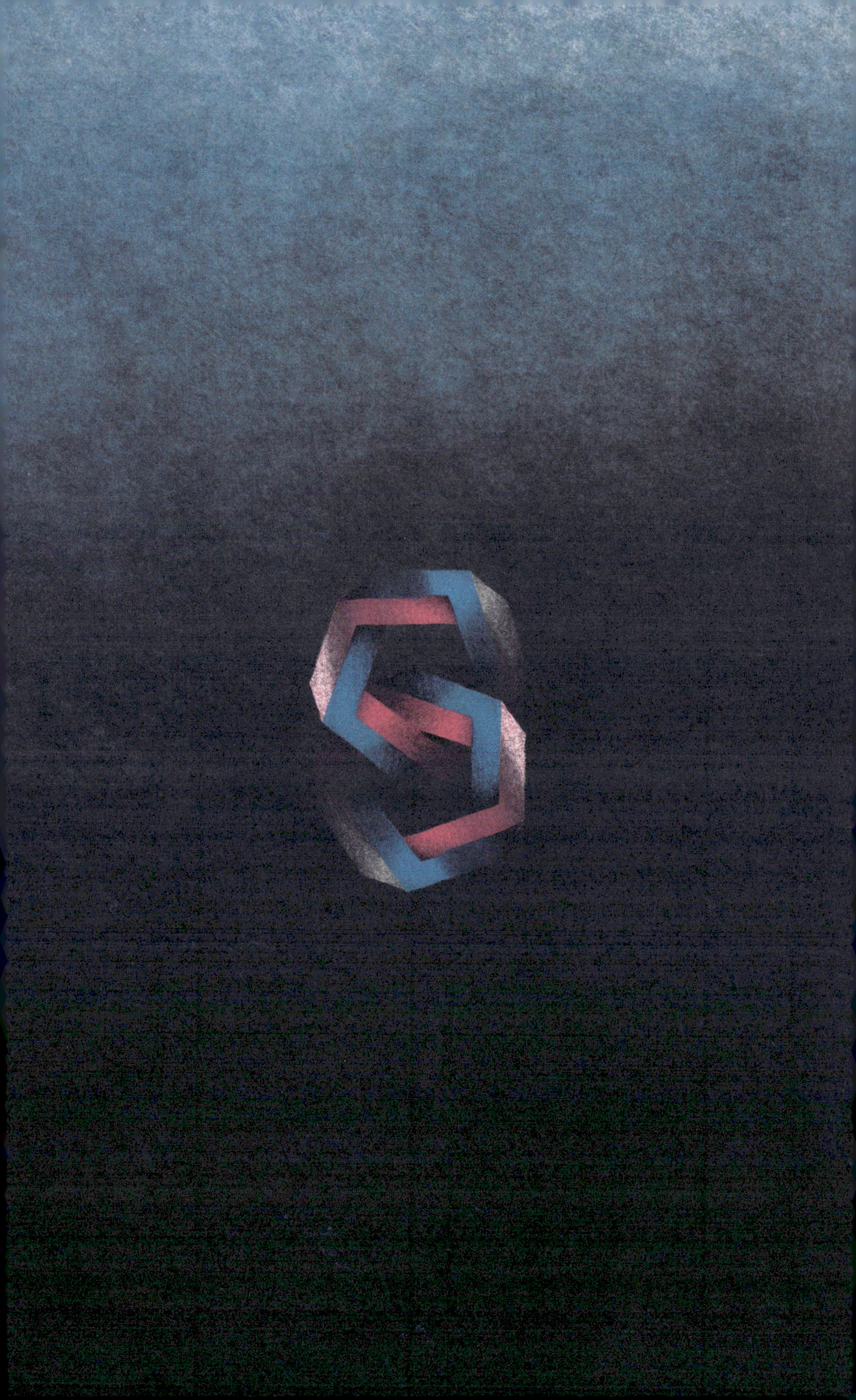

17

Reality Disappears

A paradox in the study of quantum particles leads to an astonishing and surely impossible conclusion: nothing is real.

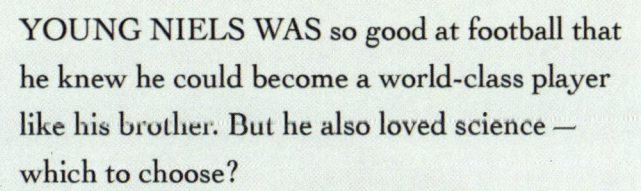

YOUNG NIELS WAS so good at football that he knew he could become a world-class player like his brother. But he also loved science — which to choose?

Tough call.

Lucky for us, Niels Bohr chose the second option, and went on to make a contribution as big (some say bigger) as Einstein in that field.

Denmark-born Bohr drifted away from both football and his family's religious beliefs when he was a teenager.

So it was ironic that he ended up bringing what some call metaphysics back to the heart of science.

This is how it happened. Scientists had argued for hundreds of years over whether light was a wave or a particle.

In the early decades of the 1900s, a series of experiments proved conclusively that it was a wave, while other experiments proved equally conclusively that it was a particle!

The scientific community was baffled.

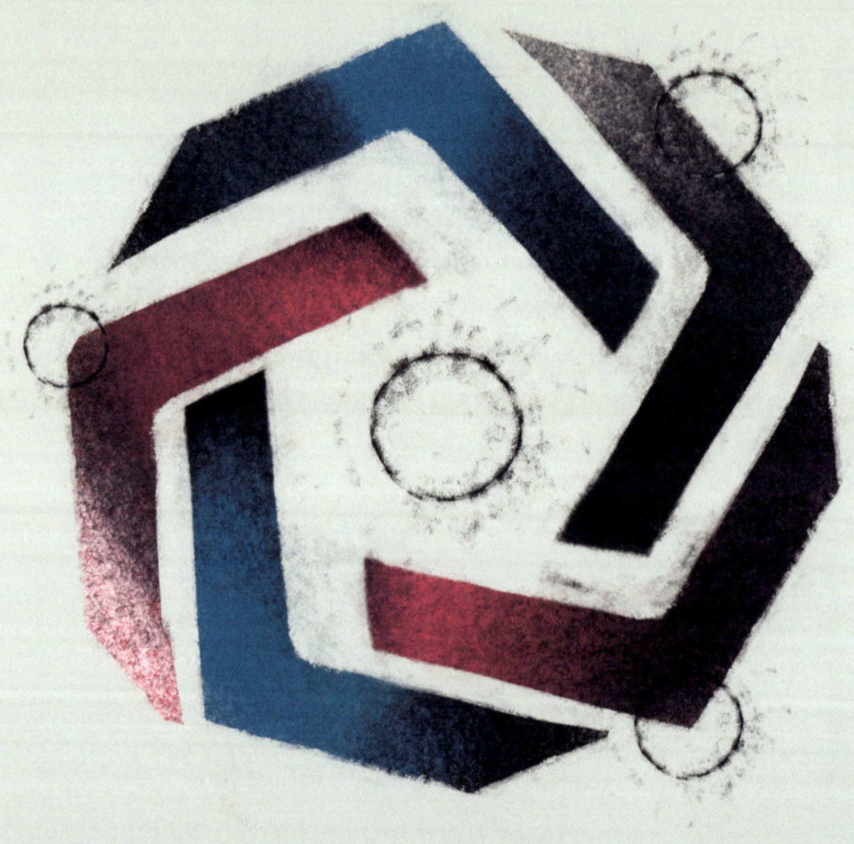

Bohr took it upon himself to learn more about this, and he felt the answer was hidden inside the atom.

Democritus's uncuttables weren't really uncuttable at all, as Ernest Rutherford had proved in our previous story. Quantum particles made up the nucleus in the middle of the atom, and also the fuzzy outer layers of each atom.

Bohr made several breakthroughs in understanding atoms before he and his colleagues made an amazing discovery.

What the experiments really showed, he realized, was that quantum particles seemed to "choose" whether to act as waves or particles depending on strange circumstances, such as whether they were being measured or not!

Huh? That made no sense. Particles don't have eyes, so how could they know whether scientists were observing them or not? And they don't have minds, so how can they make decisions?

Niels Bohr realized that the results showed that the notion of the world being "real" in the normal sense of the word was wrong: Reality was more metaphysical than it seemed.

The physical sciences didn't and couldn't explain how physical reality worked at all, and could only describe what was seen, he said.

This new understanding of the world became known as "the Copenhagen Interpretation", after the place in which scientists met to discuss it.

"Everything we call real is made of things that cannot be regarded as real," Bohr wrote. "If quantum mechanics hasn't profoundly shocked you, you haven't understood it."

One top scientist was particularly shocked. Albert Einstein disagreed with Bohr. Remember the meeting in Belgium, described in Story 1? Einstein was convinced that the world was real in a straightforward, physical sense, and the bizarre puzzle of unobserved quantum particles would soon be solved — but that didn't happen.

The mystery was about to deepen further.

18

The Mystery of Free Will

Einstein was astonished when a young German man showed how even the biggest computer in the universe could never predict the future — there was a deliberate glitch that ensured randomness.

138   21 Impossible Things

WERNER HEISENBERG, who had also been at that meeting in Belgium, sat in his office and thought about the cause-and-effect principle. Aristotle used it to look back and see the Uncaused Cause. Heisenberg realized that you could use it to tell the future.

The first particle in the universe, whatever it was, affected the second, which affected the third, which affected the fourth, and so on.

A French scientist had already pointed out that if you had a big enough brain (he imagined a supernatural being, but we can picture a giant computer), you could simply input two things: one was the velocity (which means the direction and speed) and the second was the position of every particle.

The computer could then tell you exactly where the particle would be in the following second — and the second after that, and the second after that, and so on.

If the computer was so big that it had information about every single particle in a specific place (your room, your brain, or the universe), it could use those two pieces of information to tell you everything that would happen for the rest of eternity in that space.

This meant that there is no such thing as free will. You and I do things because the atoms of our brains simply follow the cause-and-effect principle, causing us to think certain things, make certain decisions, and move in certain ways.

Werner Heisenberg, just like the ancient Chinese philosopher Mozi, felt this was the only answer that logic allowed. Yet also like him, he felt he couldn't accept it.

He was sure we were all missing something.

The Mystery of Free Will

The young German solved the problem in an unexpected way — by looking at laboratory results.

Heisenberg discovered that if you detected the position of a quantum particle, it somehow seemed to become impossible to find the direction and speed at which it was moving. If you knew how it was moving, it became impossible to detect the position of it.

The more precisely you knew one of the factors, the less precisely you knew the other.

This became known as "the Uncertainty Principle".

Einstein was puzzled to hear this. That would mean God had embedded a game of dice into the fabric of reality, he complained. It seemed impossible to believe.

But Heisenberg said the Uncertainty Principle was real, and it was important.

One of the things it did was to stop the universe and everything in it (including humans), from being an entirely predictable, boring machine.

Instead, while there are certainly elements of the universe that behave like machines, human life doesn't. We have free will and can make choices.

The Uncertainty Principle was so odd that it took a long time for people to accept it. But experiments showed that it seemed to be entirely valid.

Once again, reality-defender Einstein was baffled.

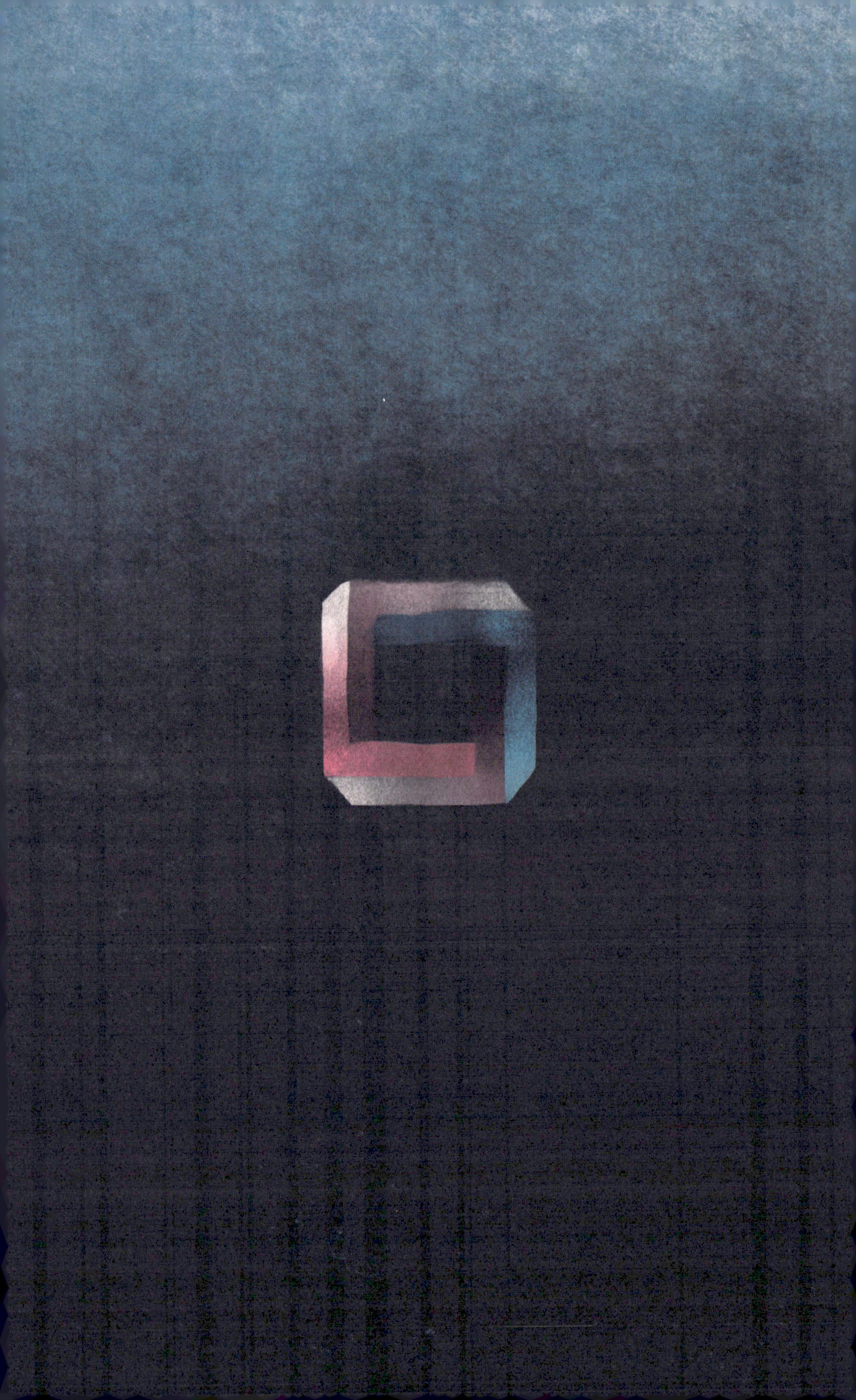

19

The Most Famous Cat in Science

A dead cat in a box is still a dead cat, right?
Maybe not.

AN AUSTRIAN MAN named Erwin Schrödinger has become famous for his cat. There are many cartoons, jokes and memes in publications and on the Internet about "Schrödinger's Cat".

This is ironic for two reasons. As far as I know, he didn't have a cat. He only made up a story about a cat, to pour scorn on quantum theory — although the cat story backfired on him and his scepticism disappeared.

Like Albert Einstein, Schrödinger found the quantum physics ideas of Niels Bohr and Werner Heisenberg hard to accept. If things only became real when we detected them, reality truly was puzzling, even "unscientific".

To show how crazy their vision was, Schrödinger said it was like imagining a cat in a large box. That cat may or may not have been killed by randomly interacting chemicals sharing the space.

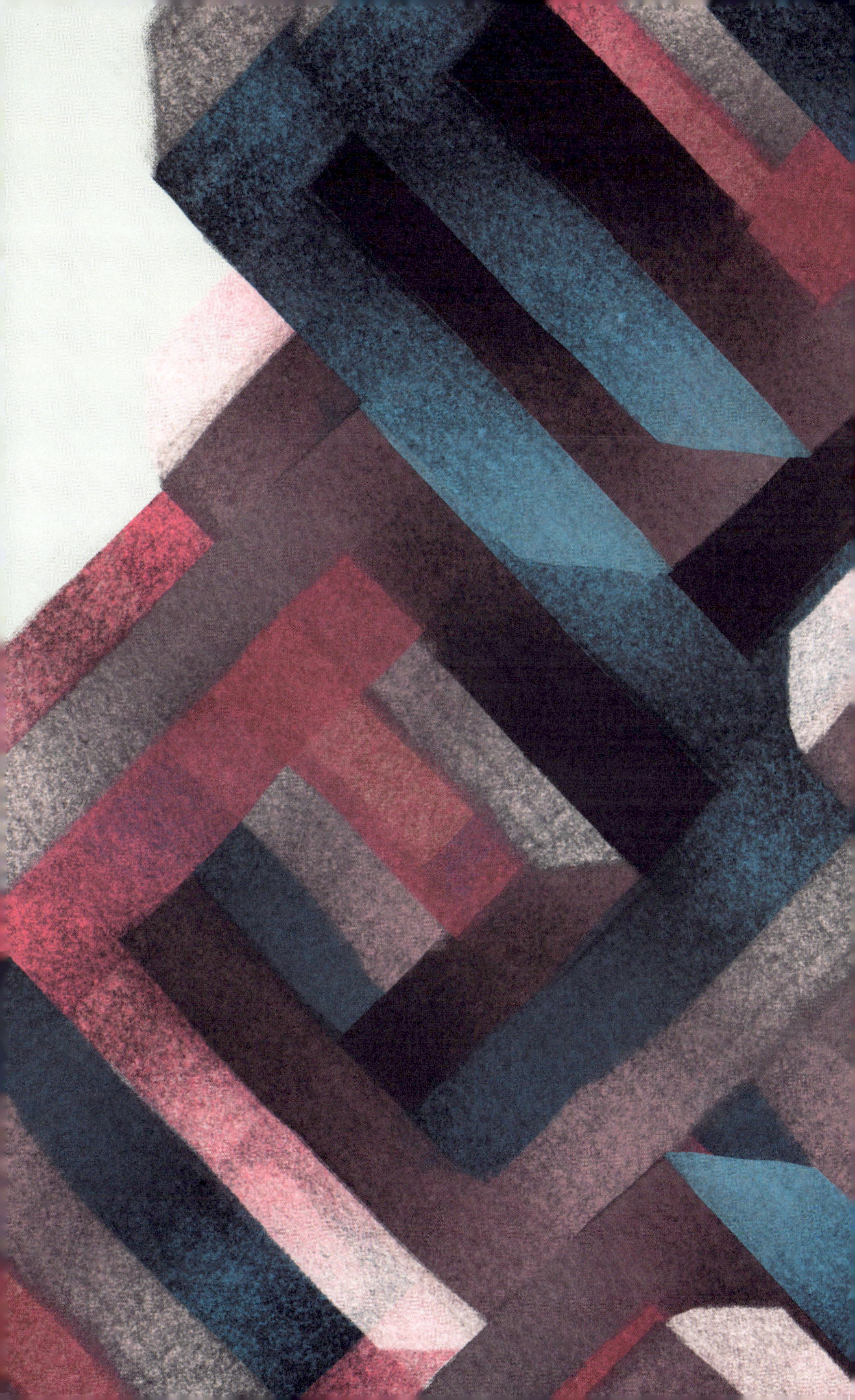

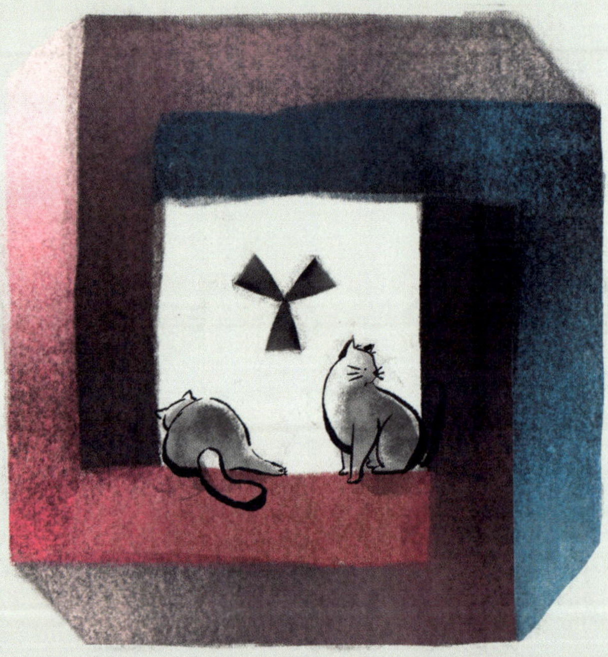

If the quantum physicists were right, the cat was neither alive nor dead until someone opened the box.

How come? Because until that moment, it was only a potential cat made up of potential particles.

When that person looked inside, he or she became the person who measured, detected or interacted with the quantum particles which made up the cat, causing them to become real.

The particles would instantly become a cat who died some time ago, or they would instantly become a cat who had remained alive throughout the whole experience.

Either way, the act of observation would create an object, complete with a past. (Think about how strange this aspect of quantum physics is — actions taken now create things that happened earlier!)

Surely this could not possibly be true.

But Schrödinger lost his opposition to quantum physics — in fact, he made a discovery that became a key part of the new science.

Now this German man was a bit of a Casanova figure, a dandy who dressed in fine clothes and had lots of love affairs.

Once, when he was having a weekend getaway in a mountain resort with a mistress, he took a break from his romantic interest to do some science in his writing book.

He surprised himself by managing to work out the exact formula that formed the bridge between quantum particles before they had been measured (when they were only potential elements), and after they were measured (when they became the actual elements that made up atoms).

It changed his view of reality.

"We do not belong to this material world that science constructs for us. We are not in it; we are outside. We are only spectators," Schrödinger afterwards wrote.

"The reason why we believe that we are in it, that we belong to the picture, is that our bodies are in the picture."

Stranger still, his imaginary not-dead-not-alive cat became the symbol of quantum physics.

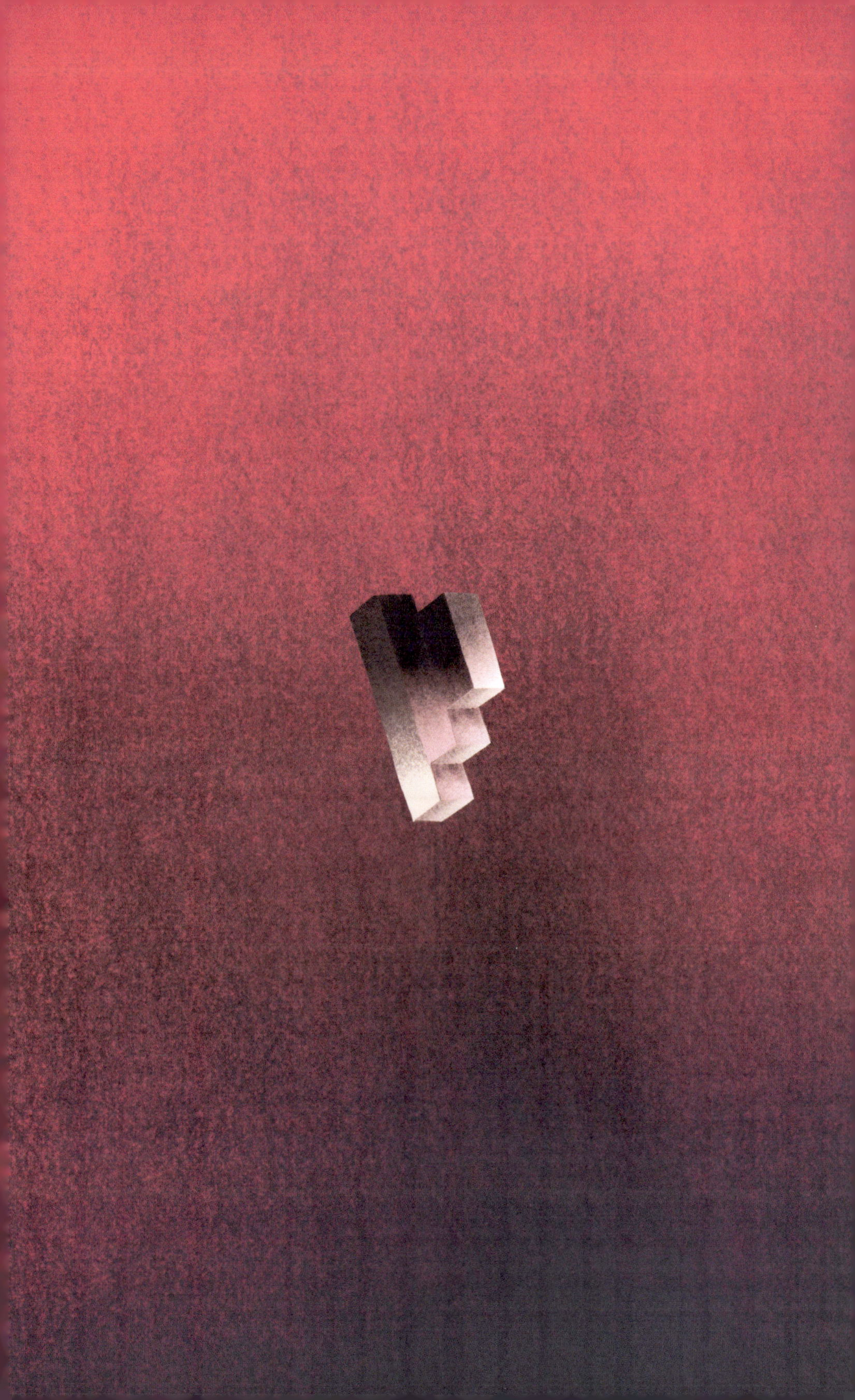

20

The Cursed Scientist

Inside each atom, something that
seems to make no sense whatsoever
is going on.

ONCE THERE WAS a scientist who seemed to be cursed. Whenever he entered a laboratory, the equipment would break.

Some scientists started to refuse to allow him to enter their labs.

But it was fine — because the man, an Austrian named Wolfgang Pauli, preferred to be a theoretical physicist. That means he didn't work in a lab. His job was to think.

It was something he did very well, often solving major science puzzles at a desk, or when out for a walk.

By the 1920s, scientists had worked out that atoms, although largely empty, contained important tiny elements in very specific places.

But there was a big problem. An atom was a large empty space and there was no structure in it to hold anything in place. What kept everything in the right spot?

Wolfgang Pauli eventually realized what the answer was: the electrons inside the atom were territorial (like guard dogs) and somehow managed the situation as if they "knew" what else was inside the atom and where it was. They wouldn't let anything overlap.

It was a crazy idea. How could these tiny items "know" anything, let alone everything else that was happening in that invisibly small world?

Pauli felt that this may even be a sort of bedrock law of nature: "No two identical electrons may occupy the same place at the same time."

Scientists in laboratories checked out this idea and found that it worked. It solved many puzzles of the structure of the atom. The "behavior" applied to the three types of particles that were key to making up ordinary matter: electrons, protons, and neutrons.

The rule, which came to be known as "the Pauli Exclusion Principle" was what made atoms, these little balls of 99.99999% nothingness, into space-occupying objects.

That's why your bottom doesn't fall through the chair and the chair doesn't fall through the floor — even though they are made of atoms, and atoms are made almost entirely out of nothing.

The only difficulty with the discovery, of course, is that no one could explain how it worked — and still can't today, a century later.

Pauli did not find this to be a problem. He came to believe that you had to be very open-minded to try to understand reality. He became friends with a man named Carl Jung, who was known as one of the fathers of psychoanalysis and believed there was such a thing as a "collective unconscious".

Pauli concluded that there were two main ways to see reality. One (which he felt was more popular in the West) saw it as a physical place for scientists to study, with physical tools such as microscopes. The other (which he felt was more popular in the East) said that it was a spiritual place which could best be explored through non-physical means, involving your consciousness.

Neither told the full story, but both were absolutely vital to humanity and were the root of positive things, he said. The first generated the physical sciences and the second generated spirituality and faith.

Many other scientists, particularly Werner Heisenberg, were impressed by his wisdom.

It was clear that Pauli's curse — to be kept out of laboratories and limited to thinking about the nature of reality — had proved to be a blessing to mankind.

21

Twins Who are a Universe Apart

Despite what your senses tell you, all objects, including humans, appear to be made of particles borrowed from another dimension.

BY THE 1930s, there had been years of tension between Albert "this reality is real" Einstein, and his puzzling and rather annoying (to him) colleagues making advances in the new science of quantum physics.

It was awkward, because Einstein was still by far the most famous scientist in the world — but the others kept winning all the disputes.

There was one more major battle to be had. In 1935, Einstein teamed up with two younger physicists, and they wrote a paper pointing out how they had found another thing in quantum physics theory that was clearly impossible.

And since it couldn't happen, it meant that the theory of quantum physics was wrong, or at the very least incomplete.

Under the rules that had been identified to govern quantum physics, you can take a tiny sub-atomic particle in its fuzzy, "potential" state, and split it into "twins". You can put one in a different room.

When you use your detector machines to measure one of them, it becomes "real", a tiny particle with a specific spin — and so does the other, at exactly the same time. However far apart they are, the two remain "entangled", to use the word of Einstein's friend Schrödinger.

Clearly, a piece of information ("I've been measured and I am now a particle which spins in this direction!") must have been transmitted from one particle to another.

Now the agreed workings of quantum physics required that this experiment must work regardless of distance. If you took one of the particles far away — say to another building, or to the other side of town, or even to another planet — it would STILL have to react instantaneously.

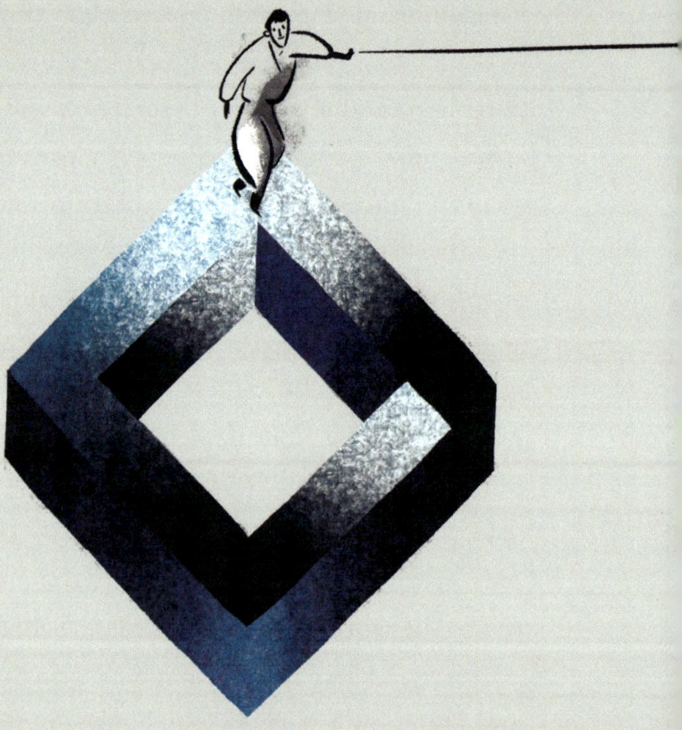

Twins Who are a Universe Apart

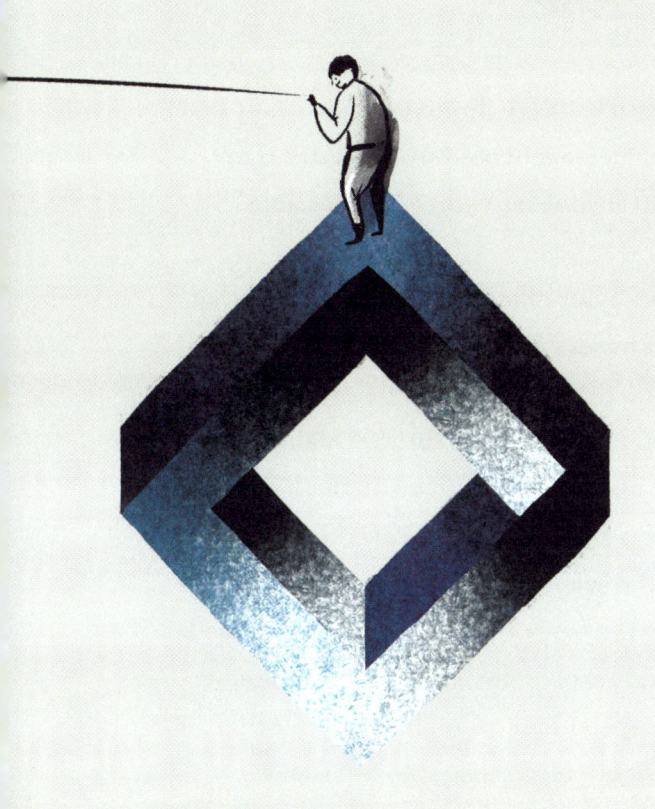

But if the particles were far apart, this meant that the mysterious message would travel faster than light — which everyone agreed was impossible!

Gotcha!

Einstein's theory of relativity had confirmed that the speed of light was actually the speed limit of reality (in Story 14). So quantum theory could not be right — or at least not right for this reality.

This paper's idea became known as the Einstein-Podolsky-Rosen Paradox (EPR Paradox), after the initials of the scientists who wrote it.

The puzzle at the heart of it was widely discussed, but no one could resolve the issue. Einstein's understanding of the speed limit of reality was clearly true. But quantum theory also seemed to be true.

There was no technology at the time that could be used to actually create entangled particle "twins" and move them to different places to test it.

Einstein died in 1955, and in the 1960s and 1970s, the physics community developed new theories and new technology for experiments — and worked on solving this particular mystery.

It has only been in recent years that we have developed the equipment that enables us to physically take a twinned pair of sub-atomic particles and separate them into different rooms or different buildings, even on opposite sides of a university campus.

When scientists actually did this test, what happened? The particles acted as if they were still joined, with the message seeming to move instantaneously, faster than the speed of light. Even though it was impossible.

Their work revealed that once again Einstein's side was wrong—and the quantum physicists were right.

Since it was impossible for messages to travel faster than light, what could this mean?

They realized that the paired particles could be any number of kilometers apart in this reality, but seemed to belong to a different dimension where they were close together, or where the concept of distances did not mean anything.

Our old friend Hermes Trismegistus (Story 6) would not have been surprised: he'd said long ago that the central belief of wise people was that there were two realities, and the one in which we lived was the less substantial one.

Entanglement of particles is a fascinating subject. It has given rise to plans for instant transportation devices, as seen in science fiction TV shows.

We shouldn't feel insecure about the fact that our reality is less real than it seems — because the ideas and possibilities that this discovery sparks are so mind-expanding.

EPILOGUE

The Journey is Just Beginning

NEITHER YOU NOR I have reached the end of the trips we are on, but we have reached the end of this brief book introducing key ideas about quantum physics and relativity. Maybe you'll be interested in reading further into one or both of these wonderful areas of science.

But before we part, here's some general advice that I hope you'll find useful.

You Use these Discoveries Already

First, we've talked a lot about mysteries, and things that seem impossible. Does that mean that quantum physics and relativity are still strange, mystical sciences? The answer is: not entirely.

Scientists tend to be practical, get-on-with-it people, and the 20th century debates about which I have written were soon overtaken by a simple phrase: "Shut up and calculate." This sounds rather rude, but it's really just science people delivering a message to ourselves that we mustn't get bogged down with mysteries when we can just go ahead and do useful stuff.

And we did do useful stuff.

The principles of quantum physics and of relativity are used widely by many people everyday — although most would have no idea that they are doing so.

For example, consider the electronics inside your smartphone, if you have one. Quantum physics principles are used to enable them to work as well as they do — the way the electricity moves through tiny, miniaturized elements takes advantage of the things that happen at particle and sub-atomic particle level.

Or consider the satellite navigation system that is built into many modern cars these days, or even the map apps in your phone. Signals are transmitted from far in the sky to your vehicle or phone – and they are only accurate because the principles of Einstein's relativity (the curious way space and time and speed affect each other) are built into the calculations. If they weren't, there would be no chance of the satellite keeping track of where you are from moment to moment.

What about the quantum theory developers' argument that reality isn't real?

Well, they did win that argument. These days, the majority of theoretical physicists accept that our natural objective understanding of reality – that we walk and talk in a world of three-dimensional objects – is not the case. It just looks that way. Yes, that is a mind-blowing thought.

Leaving out the Math

Second, I left something important out of this book. Did you notice there was hardly any mathematics in it?

That was deliberate. Many people shy away from physics because they think it involves complicated sums.

It can be done without the mathematics, as I hope this book shows. But this is only an introduction. If you are thinking of studying it at university level, there will be some serious number-crunching going on — be prepared for that. But you won't have to fill a whole giant blackboard with a single massive formula, as scientists do in Hollywood movies!

Reality and Open-Mindedness

Third, when I have given talks about these topics, I find many people find the view of reality that science at this level gives to be fresh and intriguing.

Before hearing these ideas, they expected physics to be a compilation of dry facts which remove the "enchantment" from the world — and in the hands of some TV presenters and science writers, that's sadly true.

Yet that should not be the case. In truth, science, both ancient and modern, gives us a view of reality that benefits from having an open-minded stance.

Yes, science DOES say that there is another reality, more fundamental than this one, which supports this one. Where is it and what does it look like? Your guess is as good as mine.

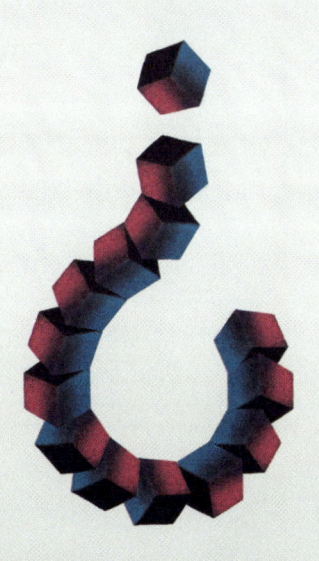

Yes, science DOES say that subatomic particles don't exist in a specific place until they are measured. So who or what measured the first particle? Was it God? Was it an alien? Was it you, travelling back in time? Maybe all three are the same thing. Or maybe everything exists at once.

Yes, science DOES say that time travel is not just possible, but happens all the time as objects move around each other at high speeds in the wider universe. If things can bend time, will we be able to do it too, one day? Or are we already doing that? There must be a way we can do that too.

Yes, science DOES say that particles create their own pasts.

What does that imply about this reality and how old it is? Or does the concept of age lose all meaning? Was the past created in the past? Perhaps we are creating it now.

These are wonderful, rich, mind-boggling ideas: and they will keep developing as we move forward and keep our minds open.

To continue following the track, read widely and avoid the extremes.

This writer is NOT a fan of those science books that were popular in the early 2000s, in which writers conjured up a fake and unhelpful war between science and religion. The truth is that spiritual faiths have preserved many of humanity's most important ancient ideas, and these should be treasured, not dismissed.

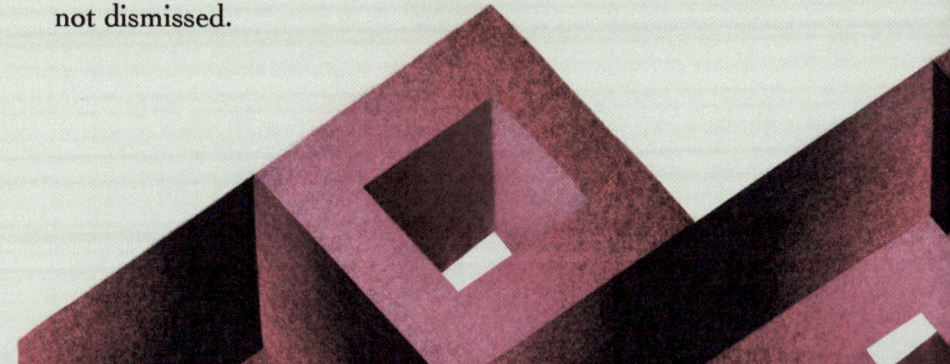

But I am also NOT a fan of pseudoscientific books which say that we create our own realities by thinking ourselves rich and beautiful. Life doesn't work like that.

Instead, look for thoughtful writers who keep their feet on the ground — but are not scared to put their heads in the clouds. *Just Six Numbers* by Martin Rees, an easy-to-read physics book about the key numbers that cause this universe to exist, is a good place to start. For people who don't mind reading a far more challenging text, *The Case Against Reality* by Donald Hoffman is a superb summary of the issues.

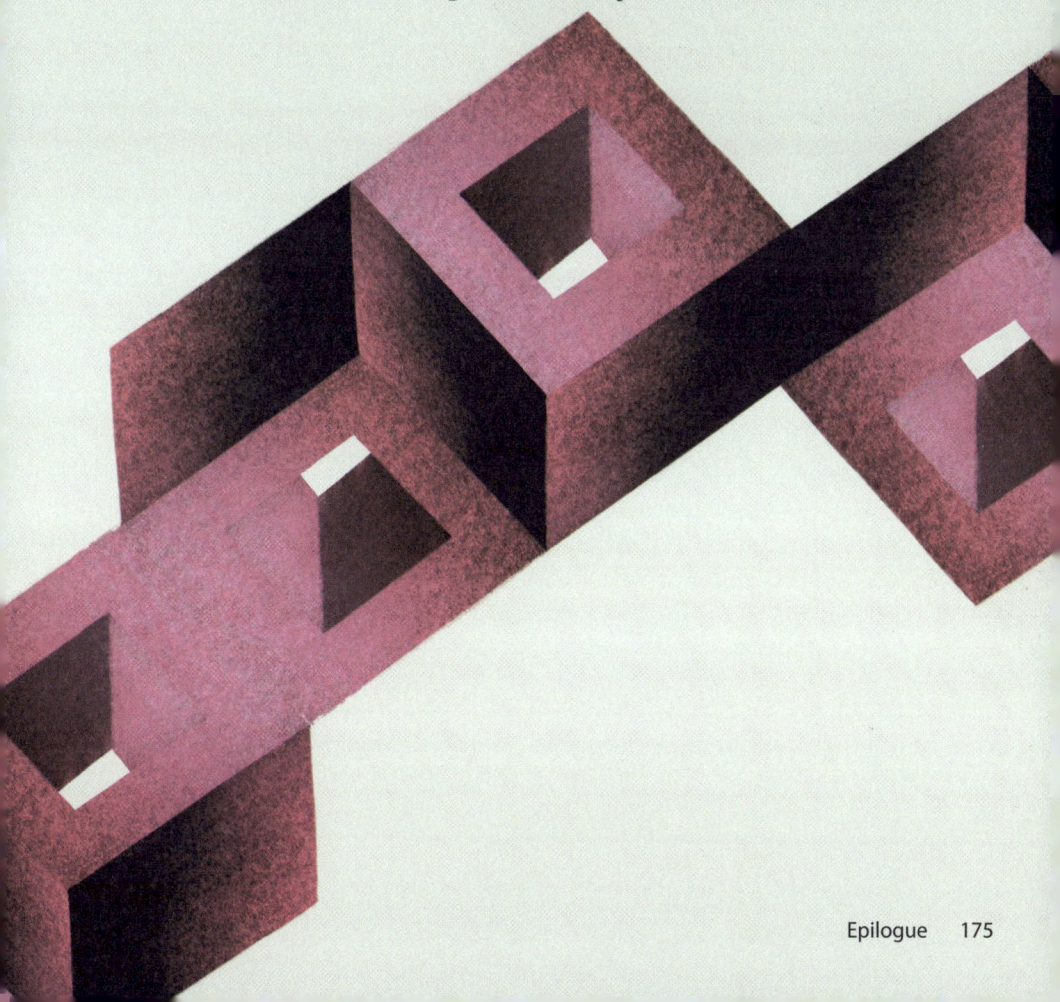

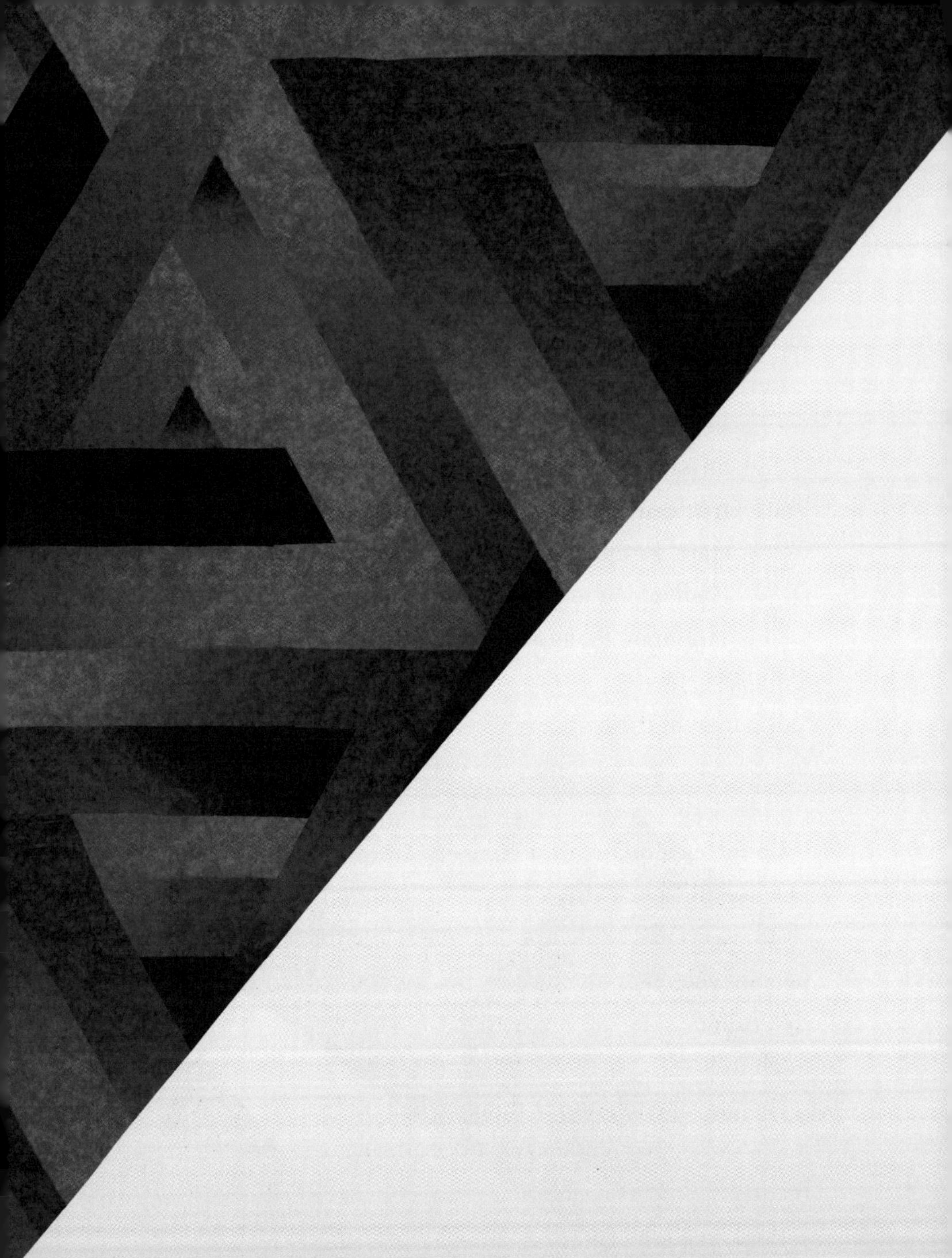

Findings

This final chapter has existed, until now, merely as a potential closing section—a fuzzy cluster of ideas which existed only in the writer's head.

But we have now firmly interacted with it, and we can see it becoming solid.

And, as so often happens in science, we find a surprise: to sum up the messages of this book, we find that we can do it best by having a bonus story: a 22nd tale in a book of 21 stories. So here's a hidden final track, as one used to find on music albums.

The man in the next story is considered one of the giants of physics, just like many of the other people in this book.

He focused on quantum physics and relativity, just like this book does.

He was remarkably open-minded, like the people who made the discoveries mentioned in this volume. He was a scientist, but also went to church.

He had a penchant for things which didn't make sense, just like you and I do, dear reader.

And he added some key new ingredients to the tale we are telling. So here's our bonus story.

22

Bonus Tale: The Do-It-Yourself Universe

It's clearly impossible for a human to use the sky as his laboratory and starlight as his experimental substance — but one physicist found a way to do just that.

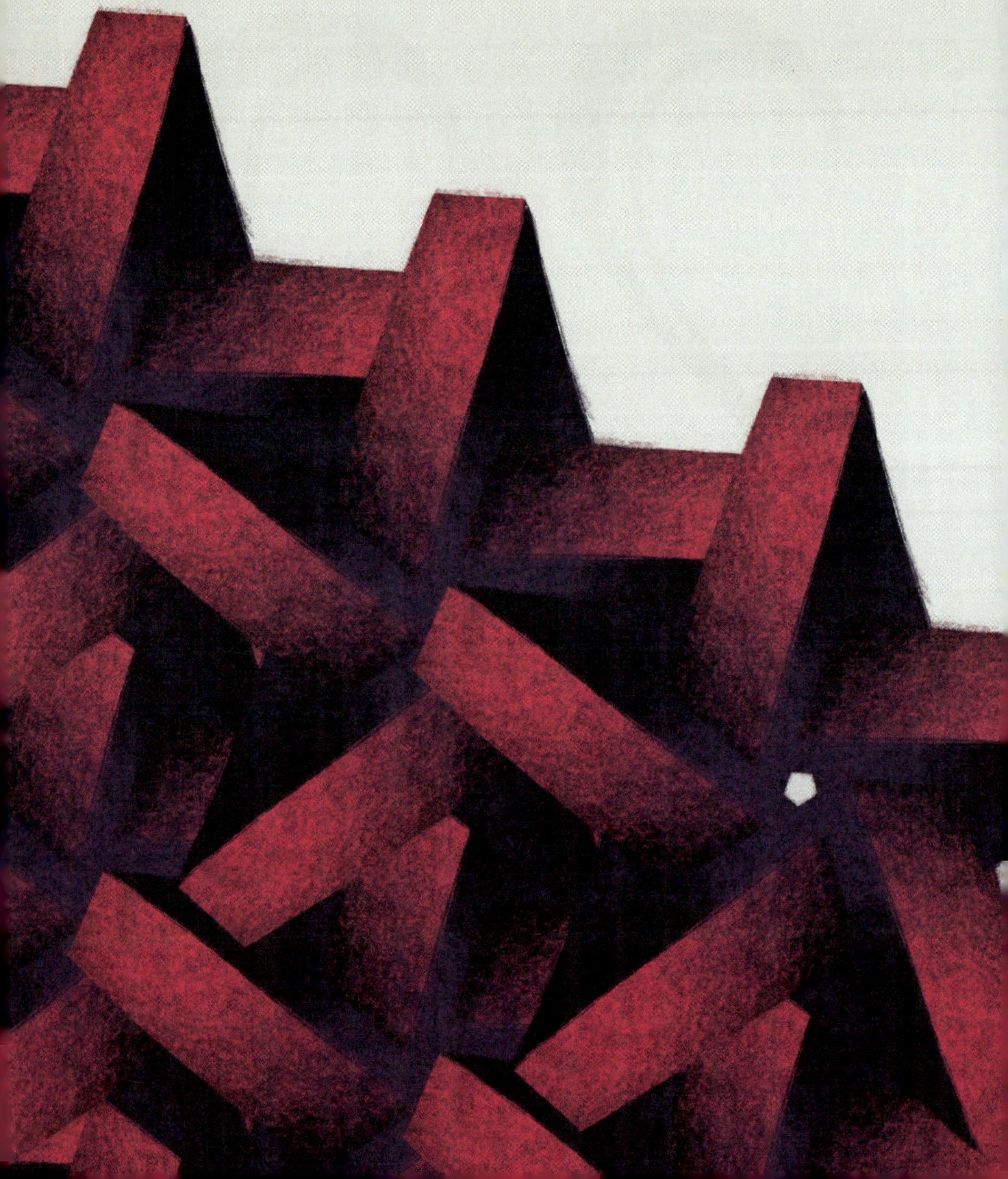

JOHN WHEELER WAS a university professor with an odd belief about education. He thought that the real purpose of universities was to enable students to teach teachers.

This is how it worked.

He would throw out thought-provoking questions. Students would try to answer them, using their natural youthful creativity. Discussions would follow, and his mind would be set off in all sorts of new directions — and so professors and students together would expand the limits of knowledge.

It was a "participatory" community. Universities function because of the "all-join-in" cooperation between the people in them, he said.

When John Wheeler wasn't teaching, he was exploring the edges of present beliefs about quantum physics and relativity, and adding much to our understanding of them.

He introduced the world to wormholes — tunnels in space and time. He also popularized the term "black hole".

But out of all the impossible things about 20th century physics, the one that seemed most intriguing to him was the way an act of observation would create an object, complete with a past.

21 Impossible Things

So that became one of his main focuses.

Scientists already knew that unwatched photons apparently behaved like waves, while observed photons behaved like particles.

Wheeler came up with what was called the "delayed choice" experiment. You send a photon on a longer-than-usual journey through machinery in a laboratory. You install separate detectors that will look at the photon near the beginning of the journey and near the end.

Pow! You trigger the machine and send the photon on its way.

But you only decide whether or not to observe it just before the very end of that journey.

Afterwards, you inspect the first detector, the one which tells you how it behaved on the earlier part of the journey.

Does your decision at the very end of the journey change the particle's behavior in the earlier part of the trip — meaning, backwards in time?

They did the experiment. And the answer was: Yes, it does.

That means we have power over the past. It was a strange and intriguing finding. The decision we took at the end of the experiment altered the nature of the particle at the start of it.

But that was just the beginning. Wheeler realized that you could do this on a huge scale, by observing photons of light coming from across the universe, some of which go past unusual gravity wells, from giant objects called quasars, that could alter them into waves and particles.

When you find one of these in the sky, you can detect the photons coming from them.

Wheeler encouraged people to look at one — and decide whether they see particles behaving like particles, or particles behaving like waves — and then told the viewers that they have been like that for their entire multi-billion year journey, thanks to what you focused on at the moment.

In other words, those photons may be 10 billion years old, yet the reason why they took that particular shape 10 billion years ago is that YOU observed them today.

It was a truly mind-boggling idea. Today's decisions create yesterday. In fact, today's decisions create billion of years of yesterdays!

Wheeler's notion of "participatory schools" turned out to be very successful. We can tell, because so many of his students became famous scientists too. Their surnames, including Feynman, Everett, Bekenstein, Wightman, Misner, and Schumacher, are all in the index listings of many physics textbooks today.

But there's something else.

One night, as Wheeler looked up at the night sky, he had an even bigger realization.

We live in a participatory universe. That's what quantum physics teaches us.

Somehow, there's a direct connection between your consciousness and the entire universe. It's a do-it-yourself universe, an all-join-in place, a reality in which everything is connected to everything else. Quantum physics is about making choices — just like the old thinkers around the world have been saying for millennia, a belief that dominates ancient faiths and cultural frameworks.

He felt that only when we understand that, do all the impossible things about quantum physics and relativity start to make sense.

On the scale of stars and space, we are tiny little nothings.

Yet at the same time, the universe is, somehow, all about us.

That is, of course, completely impossible.

And yet...

Published by

WS Education, an imprint of
World Scientific Publishing Co. Pte. Ltd.
5 Toh Tuck Link, Singapore 596224
USA office: 27 Warren Street, Suite 401-402, Hackensack, NJ 07601
UK office: 57 Shelton Street, Covent Garden, London WC2H 9HE

British Library Cataloguing-in-Publication Data
A catalogue record for this book is available from the British Library.

21 IMPOSSIBLE THINGS
Quantum Physics and Relativity for Everyone

Copyright © 2021 by World Scientific Publishing Co. Pte. Ltd.

All rights reserved. This book, or parts thereof, may not be reproduced in any form or by any means, electronic or mechanical, including photocopying, recording or any information storage and retrieval system now known or to be invented, without written permission from the publisher.

For photocopying of material in this volume, please pay a copying fee through the Copyright Clearance Center, Inc., 222 Rosewood Drive, Danvers, MA 01923, USA. In this case permission to photocopy is not required from the publisher.

ISBN 978-981-123-588-7 (paperback)
ISBN 978-981-123-589-4 (ebook for institutions)
ISBN 978-981-123-590-0 (ebook for individuals)

For any available supplementary material, please visit
https://www.worldscientific.com/worldscibooks/10.1142/12248#t=suppl

Printed in Singapore